Sex Sleep Eat Drink Dream

BOOKS BY JENNIFER ACKERMAN

Notes from the Shore

Chance in the House of Fate:
A Natural History of Heredity

Sex Sleep Eat Drink Dream:
A Day in the Life of Your Body

Sex Sleep Eat
Drink Dream

A DAY IN THE LIFE
OF YOUR BODY

Jennifer Ackerman

A MARINER BOOK
HOUGHTON MIFFLIN COMPANY
Boston • New York

First Mariner Books edition 2008

Copyright © 2007 by Jennifer Ackerman

www.houghtonmifflinbooks.com

Library of Congress Cataloging-in-Publication Data
Ackerman, Jennifer.
Sex sleep eat drink dream : a day in the life of your body /
Jennifer Ackerman.
p. cm.
Includes bibliographical references and index.
ISBN-13: 978-0-618-18758-4
ISBN-10: 0-618-18758-8
1. Body, Human — Popular works. 2. Human
physiology — Popular works. I. Title.
QP38.A155 2007
612 — dc22   2007008516

ISBN 978-0-547-08560-9 (pbk.)

Book design by Melissa Lotfy

Printed in the United States of America

DOC 10 9 8 7 6 5 4 3

To my father,
William Gorham,

WITH LOVE

Contents

Night

I stand in awe of my body, this matter to which
I am bound . . . Talk of mysteries!

HENRY DAVID THOREAU

Prologue

YOU ARE YOUR BODY. It holds you in and holds you up. It constrains you and controls you, delights and disgusts you. And yet its activities are mostly a mystery. Let's face it: We're all body-conscious to one degree or another, acutely aware of our physical façades— the symmetry and wrinkle of face, the curve of torso, girth of thigh, roll of belly, flare of feet. But how many of us have a handle on the drama unfolding inside? As Saint Augustine said, we go forth to wonder at the heights of mountains and the courses of the stars, yet pass by the miracle of our own inner lives without wondering. In health, the body often operates so smoothly that we can almost forget it exists. Most often it's some failure or perturbation that captures the attention. In fact, many of us spend our time trying *not* to be aware of what's occurring within. No news is good news.

Not so. This came home to me some time ago when I succumbed to a virulent flu after a stressful run of life. The flu sucked the juice out of me for weeks and robbed me of all the facets of physical existence I relish: the satisfactions of work and exercise, the sweet smell of my children and other sensual pleasures, appetite and eating, restful sleep. When I emerged from my illness, I felt not only the relief and joy of having my body back, but a sudden sharp desire to learn more about it. What was the nature of those pleasures my healthy body so enjoys? And the problems that occasionally plague it? I realized I didn't have a clue what went on inside me, in sickness or in health. I had no idea, for example, what underlies digestion and its precursor, hunger—that mysterious loop that translates the absence of nutrients into the craving for comfort food—or, for that matter, its antithesis, nausea. I hadn't the foggiest idea what a virus did to my body, or alcohol to my brain, or cumulative stress to my energy and health. I knew my body did some things more effectively in the morning, others in the afternoon or evening, but didn't have an inkling why.

Though that bout of flu was hardly a near-death experience, it did remind me that my whole existence was going to come and go in this same ark of skin and blood and bones; the "go" part, of course, loomed closer every day. Even the long-lived among us are alive for only about 700,000 hours. My body would exist only once; I would never have another one. Wouldn't it be a good idea to get to know it a little?

When I was in first grade, I had a fine grasp of my inner life. I knew that my heart beat somewhere inside my left chest, near where I put my hand to pledge allegiance. I knew that when I brushed my hair, I was stroking dead cells, a grotesquery I gleefully shared with friends at every opportunity. I knew that what my stomach took in as a snack—say, a whole box of raisins—might have later consequences. I knew I would get cranky if I didn't have a nap. Beyond that, I didn't give it much thought. This went on, more or less, for thirty years. Then came that flu that struck like lightning on the road to Damascus.

To remedy my ignorance, my first thought was medical school. I imagined poring through *Gray's Anatomy*, committing to memory nerve and bone, perusing the *Lancet* and the *New England Journal of Medicine* for case studies describing mysterious clinical syndromes: "A 10-year-old girl with recurrent bouts of abdominal pain," or "A 22-year-old man with chills and fever after a stay in South America." Medicine had the appeal of detective work: observing closely, analyzing, diagnosing, offering treatment. But starting a medical education from scratch at the age of thirty-five would rule out normal life well through my childbearing years.

Also, I did know one thing about my body. It lacks the constitutional prerequisite for the kind of schedule demanded of doctors: It needs sleep. The night before committing myself to a two-year post-baccalaureate premedical program, I dreamed of diving off a bridge and landing headfirst in a slough of mud. In the morning, I canceled my medical school plans.

It was another decade before I got around to tackling the topic as a writer. Over the next several years, I hunted everywhere I could for the latest engaging news about the body. I read dozens of books and hundreds of journals. I prowled the laboratories of scientists and attended their conferences, meetings, and lectures. I observed significant events in my own body and subjected it to numerous tests and experiments.

I discovered that it was a good thing I waited as long as I did. Much of what we know about the body we've learned only recently from an explosion of new discoveries. In the past five or ten years, science has made a great leap forward in grasping the underpinnings of everything from hunger and fatigue to exercise, perception, sex, sleep, even humor. We know things about the body that were scarcely imaginable a decade ago—exactly which brain regions are active as you read this sentence, for instance, or what cumulative stress does to your waistline, or how exercise can help you learn. This fresh news suggests answers to questions that once seemed beyond the reach of science: Why do you succumb to a cold and your partner does not, even though you were both exposed to the same sick child? Is there a biological basis for spousal arguments over whether those red pants match that crimson shirt? How is it that your colleague can eat anything she likes and never gain an ounce, while you just look at a doughnut and put on half a pound?

In the past decade, we have learned that the human body is only 1 percent human and 99 percent microbial, at least in terms of cell count. (That you and I don't *look* more germ-like is due to the small size of bacterial cells relative to our own.) We know that just thinking about exercise may increase muscle strength, and that too little sleep can lead to too much weight gain. We have begun to see that "timing is everything"—that if you want your body to go through life at its best, you should pay close attention not just to what you are doing, but when you are doing it.

Some of what we've learned has come from studying cases in which normal bodily functions have failed. As the seventeenth-century English anatomist Thomas Willis said, "Nature is nowhere accustomed more openly to display her secret mysteries than in cases where she shows traces of her workings apart from the beaten path." From appetite gone awry we've glimpsed the chemical essence of hunger. From a failure to recognize faces we've garnered new insights into the miracle of face perception; from one who "lost touch" we've learned about the biology of a caress.

Other scientific breakthroughs have arisen from innovative tools for seeing inside the body. In centuries past, studies have required a bizarre injury to expose the previously concealed innards of an unfortunate patient. The closest thing we had to a real window on the workings of an organ was accidental—a chance hole in the stomach of Alexis St. Martin, for example, which gave an army doctor named William Beaumont an intimate view of the digestive organ at work. This was followed by the first x-ray photographs in the twentieth century, which yielded clear but static images of bones in their misty envelope of flesh. In the past ten or twenty years, new imaging techniques—positron emission tomography (PET) scans and functional magnetic resonance imaging (fMRI)—and ways of "listening in" on the activities of cells have allowed detailed looks inside the living, working body. Brain scans have pointed a bright spotlight at what's happening in the brain in real time as we recognize a face, learn a new language, find our way around a Byzantine city, follow a Bach sonata, or get a joke. With tools that allow us to eavesdrop on the cells of the human gut, we've discovered the existence of a "second brain" there, as well as a world of organisms living in its twisted topology of villi and crypts.

So, too, huge strides in genetics have helped us explore in a whole new way the fundamental workings of organs, tissues, and cells. The lion's share of new knowledge on human genes has been gleaned from the study of other organisms: mice, fruit flies, zebra fish. Much to the delight of scientists, the mechanisms that run creatures from fungi to humans often have a common basis. What is true of lowly yeast is also true of you.

Among the fascinating new findings is this: An essential part of our inner life is rhythmic. "Our body is like a clock," wrote the scholar Robert Burton in 1621. It's true. We are not just time-minded but time-bodied, right down to our very core. The human body possesses a whole shop of internal clocks that measure out our lives. These timekeepers tick away in a "master" clock in the brain and in the individual cells throughout our flesh, affecting everything from the time we prefer to wake up in the morning to the accuracy of our afternoon proofreading, our speed during an evening run, even the strength of our handshake at a late-night party. We are usually unaware of the internal rhythms generated by these clocks, sensing them vividly only when we abuse them, during shift work, jet lag, or adjustment to day-light-savings time. Yet they govern the daily fluctuations of a surprising range of bodily tasks, from the operation of individual genes right on up to complex behaviors—how we perform in sports, tolerate alcohol, respond to cognitive challenges. By timing your actions so they're in concert with these rhythms, you can maximize your performance at a meeting or minimize your dental pain. By defying them, you may cause yourself real harm.

This is a book about the new science of your body, the many intricate and intriguing events occurring inside it over a twenty-four-hour day. There is, of course, no typical day. Nor is there a typical body experience. (In using the first person, I am borrowing a tack from Thoreau: "I should not talk so much about myself if there were anybody else whom I knew as well.") Physicists may deal in uniformity, in things that are all the same, such as electrons and water molecules. But biologists must cope with staggering diversity. No two animals are alike, even when they're clones. The same is true for two cells and two molecules of DNA. And while recent research suggests that we humans are genetically more alike than we are different, we are nevertheless marked by millions of small but significant distinctions of anatomy, physiology, and behavior. We diverge in our appetites and metabolism and in the way we taste and see. We differ in how we tolerate stress and process alcohol and in our preferences for bedtime and waking time. One man's tonic is another's poison. One woman's stimulus is another's trauma. One body's night is another's dawn.

Even within an individual, variation reigns. Over the course of a day, a year, a lifetime, we are many different people. As Montaigne said, there is as much difference between us and ourselves as between us and others.

Nevertheless, we all share common body experiences. A single volume can't hope to cover them all, or even those that transpire within the confines of a single day. The choice of topics here reflects my own preoccupations, as well as guesses about what will prove interesting for others. From caress to orgasm, multitasking to memorizing, working out to stressing out, drooping to dreaming, it's here.

Morning

The brightness of a new page
where everything yet can happen.

RAINER MARIA RILKE
Book of Hours

1

AROUSAL

M Y EYES OPEN just long enough to fix on the clock: 5:28 A.M., two minutes before the alarm rings. Except for the distant fluty notes of a songbird, the world is silent. Though the stars are fading, it will be another hour before the sun's first rays creep over the horizon.

Maybe you're like me: You anticipate your alarm, wake a minute or two before it blares. It's probably not sufficient sleep that awoke you. What did, then? Some people claim that subtle aural trigger cues do the trick, those characteristic sounds of early morning, such as the start-up of noise on a highway or the passing of a delivery truck or even the little tick produced by a mechanical alarm clock just before it rings. It's true that the brain is good at processing sound while we sleep; that's why we buy audible alarm clocks. We don't buy odor alarms for equally good reason. Though some people swear they are roused from deep sleep by the putrid stink of skunk or the heady aroma of percolating coffee, a new study suggests otherwise: Scientists at Brown University documented a complete failure of response during all but the earliest phase of sleep to powerful odors such as peppermint and the distinctly noxious pyridine, a component of coal tar often used as a herbicide for firewood. Don't count on the nose as a sentinel system, say the researchers: "Human olfaction is not reliably capable of alerting a sleeper."

In any case, there's mounting evidence to suggest that the trigger cues may not be outside your body at all, but inside, in a kind of brilliant little mind-based alarm clock that prepares the brain for waking. When Peretz Lavie, a sleep researcher at the Technion–Israel Institute of Technology, investigated whether people can awaken on their own accord at a specified time without external cues, he discovered a surprise: Many of his subjects awoke ten minutes before or after the appointed time, even if it was as early as 3:30 A.M. This is a truly remarkable feat of time-telling, which probably exceeds the ability of most people to tell time during their waking hours. Another study showed that the mere expectation that sleep will come to an end at a certain time boosts by 30 percent blood concentrations of the stress hormone adrenocorticotropin (ACTH), a sure sign that the brain is gearing up for waking.

In some of us, at least, the unconscious mind somehow keeps careful track of clock time even as it sleeps, so that the brain "expects" a timed event, such as a target wake-up time, just as it does during wakefulness, and triggers the release of chemicals designed to get us up and going. Anticipation—once thought to be an ability only of the conscious mind—may actually occur as we sleep, allowing (or dooming) us to rouse spontaneously at the same predictable hour.

Talk about mysteries!

But maybe you don't have this problem; maybe you're among the majority who awaken with a startle to the real sound of a real alarm or a burst of music or DJ chatter from your clock radio. For you, the morning ritual begins with a poke at the snooze button to steal ten more minutes of sleep. Chances are you need this—and more. In a nation that averages less than seven hours of sleep instead of the optimal eight, most people are mildly sleep deprived, especially during the work week. Unfortunately, the short bouts of sleep you nip between slaps at your clock are not restorative or restful, say specialists, but light and fragmented. Even if you doze through the next sounding of the buzzer, your expectation of awakening will affect the quality of your slumber.

There are, of course, those who will stubbornly sleep through even the shrillest alarm. For such dyed-in-the-wool dozers a patent was granted in 1855 for an ejector bed. If the snoring sinner ignored a built-in alarm, the side rail released, tilting the bed so that the slothful occupant tumbled to the floor. Only slightly more humane is an apparatus newly devised by a clever crew at the Massachusetts Institute of Technology: "Clocky," a fuzzy, spongy robotic alarm clock, rolls off the bedside table and zips away on a set of wheels to some elusive corner of the room. It finds a new hiding spot every day. The arduous act of finding Clocky, say the inventors, should prevent even the sleepiest owner from revisiting the pillow.

Oh, to lie for a minute in that borderland of wake and sleep known as the hypnopompic state (from the Greek for "sleep" and "sending away"), to let the mind drift into wakefulness and relish the lovely, slow coming on of day. Few of us have this luxury. If waking feels strenuous, that's because it is; with rising comes brief but violent shifts in heart rate and blood pressure and a peak in blood levels of the stress hormone cortisol.

Alertness follows only slowly. The grogginess and disorientation immediately after awakening is known as sleep inertia, and nearly everyone suffers from it. "The brain just doesn't go from 0 to 60 in seven seconds," quips Charles Czeisler, a rhythms researcher at Harvard University. Most of us perform poorly on mental and physical tasks at daybreak compared with how we do just before retiring at night. "It's ironic," says Czeisler, "but the brain's performance in the first half hour after waking is worse than it is if you've been up for twenty-four hours." This useful piece of information was discovered the hard way by the U.S. Air Force in the 1950s. It had put in place a practice of sending pilots out to their jets on the tarmac overnight so they could sleep in their cockpits and be ready to go at an instant. The pilots were roused from sleep and told to take off; the accident rate soared, and the practice was banned.

When a team of scientists formally quantified the effects of sleep inertia in 2006, they found that the cognitive skills of test subjects were, on awakening, at least as bad as if the subjects had been legally drunk. While the worst sleep inertia is dispelled after about ten minutes, its effects may linger for up to two hours. Its severity will depend in part on the stage of sleep from which you were roused. Lavie's team found that people waking out of the stage known as rapid eye movement (REM) sleep can quickly orient themselves in their surroundings and tend to be more mentally nimble and chatty. REM sleep is a kind of portal to wakefulness, says Lavie, best smoothing the transition out of sleep. (It is also marked by intense and vivid dreaming, which may account for the fresh, lucid memory of some dreams on waking.)

On the other hand, those who are unfortunate enough to be cat-apulted into consciousness from deep, non-REM sleep by the jarring ring of an alarm are apt to be disoriented, with that "where am I?" feeling. To eliminate such rude awakenings, Axon Sleep Research Laboratories has come up with a kinder doppelgänger to Clocky called SleepSmart, which monitors your sleep cycle and awakens you out of your lighter, REM sleep phase. A headband described as "minimal, comfortable, and sleek" is fitted with electrodes and a microprocessor, which measure your brain waves during each phase of your sleep and transmit the information to a clock near your bed programmed with your latest possible wake-up time. The clock takes care of the rest, awakening you during the last light-sleep phase before the zero hour.

Whether you hop or drag into full morning alertness also hinges on your chronotype, a kind of avian profile that describes your rhythmic nature—larkish or more like an owl. Lark chronotypes emit their music at sunrise; owls side with the night.

I once heard the writer Jean Auel say that her brain works best long after sunset. She goes to work at eleven or twelve at night, finishes up at seven in the morning, and then retires. She sleeps until four in the afternoon, when she rises and eats with her husband—his dinner, her breakfast—goes out on the town, and finally settles down to work again at midnight. She claims this extreme owlish "shadow" life takes no toll.

Such is the pattern, too, for the great geneticist Seymour Benzer, whose often nocturnal studies of mutant fruit flies helped unravel the genetic basis of our daily body rhythms. Benzer's working day is the middle of the night; he says he risks accident if he's forced to start work when most people do, in the morning.

At the other end of the spectrum from Auel and Benzer are extreme larks, those partial to the bright work of bakeries, who fall asleep as early as 7 or 8 P.M. and awaken full of verve at 3 or 4 A.M. The two extreme chronotypes can seem as different from each other as people born in different centuries or on different sides of the planet, the larks stirring just as the owls are falling asleep. The "birds" differ dramatically in peak alertness (11 A.M. for larks, 3 P.M. for owls), heart rate (11 A.M. for larks, 6 P.M. for owls), and in favorite mealtime, favorite exercise time, and daily caffeine use (cups for larks, pots for owls).

Till Roenneberg, a chronobiologist at the University of Munich, has found that extreme owl types are three times more common than extreme larks. Most people fall somewhere in between, with a leaning toward mild to moderate owlishness—a pattern that often fails to fit well with work routines, leading to feelings of "social jetlag." You can assess your own status by taking a simple questionnaire devised by Roenneberg's team, which asks such questions as: What time do you normally awaken on workdays? On free days? When do you feel fully awake? At what hour do you have an energy dip?

It should be noted here that despite the many proverbs praising the virtues of larks (Benjamin Franklin's "Early to bed, early to rise," "The early bird gets the worm," and so on), science suggests that there is no health or monetary advantage to being an early riser, nor is it necessarily a sign of mental well-being. Some time ago, a group of British researchers set out to substantiate Franklin's gnomic wisdom using data collected from more than 1,200 elderly men and women. But after examining the effects of bedtimes and waking times on health, material circumstance, and cognitive function, the researchers found that owls were in fact more often wealthier than larks, and there was little difference in the health and intelligence of the two.

In any case, you may have little choice in which bird you resemble. The daily habits of larks and owls are not a result of differences in personality, as once believed, but in the nature of our internal biological clock. Almost a decade ago, Hans Van Dongen of the University of Pennsylvania demonstrated that the biological clock of average morning types is more "phase advanced" than the clock of evening types—that is, it runs earlier, by as much as two hours. Although you might be able to overcome your proclivity, says Van Dongen, you probably can't change it. Your larkishness or owlishness is likely built right into your biology.

"Time is the substance I am made of," wrote the Argentine novelist Jorge Luis Borges. There's a deep hunch here. As biologists have learned in the past decade, time permeates the flesh of all living things—and for one powerful reason: We evolved on a rotating planet.

To understand this, think back billions of years, to an earlier world where all organisms are single-celled and floating in some warm, primitive sea. The bright sun of midday cycles with the dark cool of night, day after day, periodically, predictably, for trillions of days. Light and dark, warmth and cold: In the matrix of these daily ups and downs, ins and outs, life unfolds. With no screen of ozone in place, ultraviolet radiation damaging to life bombards the earth's surface during daylight. To avoid the harmful rays, organisms limit certain fragile or sensitive biochemical processes to the dark asylum of night, generating a rhythmic metabolism. Some evolve sensors to discern the occurrence of sunlight, at first mere light-sensitive cells, and later, sophisticated eyes that help them detect the subtle transitions of dawn and dusk.

Then comes the genius. Some life forms develop genes, cells, and bodily systems capable of generating their own internal daily rhythms beautifully attuned to planetary time, known as circadian rhythms (from the Latin *circa*, about, and *dies*, day). Pathways evolve from their light sensors to these circadian clockworks to help synchronize the internal rhythms with the solar day. "In this way," says the biologist Thomas Wehr, "the circadian pacemaker creates a day and night within the organism that mirrors the world outside."

So sensitive are these pacemakers to light that even low illumination adjusts and resets them. Sunlight is their dominant *zeitgeber*, or time giver; it sets their rhythms so they remain in tune with—or entrained by—shifting patterns of daylight and darkness, so that in summer, biological day is long, and in winter, it is short. When you pull up the shade in the morning, special light-sensitive cells in your retina measure the brightness and register dawn in the dark cradle of your brain, aligning your circadian clock to cosmic rhythms.

Yet so robust and reliable are the pacemaker's rhythms that they run continuously and persist even in the absence of environmental cues. Science has discovered this through studies in which subjects are isolated from environmental cues for weeks at a time. With no clue to the turn of day and night, their bodies start to decouple from the solar cycle but adhere to a twenty-four-hour cycle of waking and sleeping and other body rhythms. (These persistent daily patterns are known as free-running rhythms and are hard-wired into a species' genome.)

The new system offers two great advantages: doing the right thing at the right time within the body, and also anticipating daily transitions and tailoring behavior in the environment accordingly. By carrying inside this model of the cosmos, the body keeps a step ahead of the changes going on around it, preparing for food, mates, predators, and temperature extremes brought on by day and night.

• • •

"Clock" seems too feeble a term for this potent circadian influence on our bodies. Though pressures are powerful to keep body conditions constant, the circadian drive causes dramatic fluctuations throughout the twenty-four-hour day. As Emerson wrote, everything looks stable until its secret is known.

Consider body temperature.

Maybe you've now popped into the shower. To wake up and get going, some people recommend running a "contrast" shower, with hot water, then cold. (This technique may do double duty, rousing others with the whooping that accompanies the cold phase.) The heat receptors just beneath your skin's surface detect temperatures up to about 113° F; the cold receptors, down to about 50°. Above or below these temperatures, pain receptors kick in. But even if you let the water get very hot or very cold, your core body temperature will not change much. (Incidentally, the number for the normal, average body temperature so well known to most of us—98.6°—is wrong. Meticulous studies involving millions of measurements have revealed that the true average daily temperature for women is 98.4°; for men, 98.1°) So skilled is the body at keeping its temperature relatively constant despite changes in the environment that a champion cold-water swimmer such as Lynne Cox can sustain her body heat in the freezing seas of the Antarctic, and a marathon runner can keep his cool in the 120° heat of Death Valley.

Our knack for holding steady our temperature and other internal conditions—called homeostasis, from the Greek words for "similar" and "steady"—may be taken for granted, but it's a remarkable phenomenon. The body maintains its internal milieu by constantly monitoring levels of glucose, carbon dioxide, hormones, temperature, even the pH of spinal fluid. These levels flutter about a set-point, or norm. An intricate and diverse network of nerves and hormones in the body senses divergences from these set-points and rectifies them by sending word to appropriate systems that can set in motion corrective mechanisms.

What we've learned lately, however, is that our set-points aren't set at all; they actually cycle in a circadian rhythm, varying according to time of day—with profound implications for how we function and the way we feel. Body temperature, for instance, typically ranges a couple of degrees Fahrenheit over the course of a day, starting out at a low of about 97° in the very early hours (so a temperature of 98.6°

first thing in the morning is, in fact, a low-grade fever) and rising to as high as 99° or even 100° in the late afternoon and early evening. These temperature swings affect all sorts of bodily experiences: When our daily temperature peaks, for example, so does our tolerance of pain and our muscle flexibility, speed of reflexes, eye-hand coordination, and proofreading accuracy.

Heart rate and blood pressure also vary with time of day, along with the number of circulating white blood cells, levels of hormones and neurotransmitters, even the velocity of blood flow in the brain. Heart rate and blood pressure slowly increase over the day; the stress hormone cortisol ebbs. With the onset of night come surges in the "darkness hormone" melatonin, a gradual fall in temperature, heart rate, and blood pressure, and a slow rise in cortisol until its peak in early morning, before waking.

These circadian oscillations are hardly a trifling matter. If physicians don't take them into account, the measurements in a given individual of everything from blood pressure to heart rate, sperm count to allergic reactions, can be badly distorted. (Some scientists even argue that every clinical observation should be "time-stamped.") The rest of us can use our knowledge of these bodily ups and downs to make good personal choices. To avoid excessive bleeding, it's best to shave at 8 A.M., when clot-forming blood platelets are more abundant and stickier than they are at other times of day (which also explains why heart attacks peak at this time). To escape those wincing twinges in the dentist's chair, time your visit for the afternoon, when the pain threshold in teeth is highest. If you want to minimize the damage to your body from alcohol, drink that beer or glass of wine between 5 and 6 P.M., when the liver is generally most efficient at detoxifying booze. And if you want to set an athletic record, schedule your race for late afternoon or early evening.

So pervasive is the influence of our circadian cycles, says the chronobiologist Josephine Arendt, "that it would be reasonable to say that everything that happens in our bodies is rhythmic until proved otherwise."

So where is our little ticker? Pop into the bathroom for a minute and look in the mirror. If you could see into the dark interior of your skull, you might catch sight of a pair of tiny wing-shaped structures in the brain's hypothalamus, just behind and below your eyes, one in the right hemisphere and one in the left. These clusters of ten thousand neurons, collectively known as the suprachiasmatic nucleus (SCN), comprise the master clock in your brain. The SCN measures the passage of a twenty-four-hour day by producing and using special proteins in a circadian pattern. It controls and organizes the big rhythms of the body so that its sleep functions are optimal at night and its wakeful functions during the day. (When the SCN is destroyed in laboratory animals, their activities—running, eating, drinking, sleeping—follow no normal twenty-four-hour pattern but are randomly distributed across the day.)

With a full-length mirror and a little genetic engineering, you might also see the rest of your body ticking. We now know that the body has not one clock but billions: Circadian timekeepers are ticking away in virtually every bit of flesh, in kidney, liver, and heart, in blood and bone and eye. In one 2004 study, researchers inserted a gene for luciferase, a protein that gives fireflies their luminescent glow, to show in real time the circadian rhythms of cells in peripheral tissues. There they were, cells from all corners of the body, "blinking" in a circadian beat.

Though the master clock in the SCN oversees the body's cyclical rhythms, the genetic timepieces pocketed in the cells of outlying tissues and organs may follow their own daily routines, triggering peaks and troughs of activity at different times of day in their respective locations to ensure that a particular organ has what it needs when it needs it, and timing its activities according to its own priorities. The clocks in heart cells, for instance, set their own daily rhythms for blood pressure, and the clocks in liver cells, for digestion and for metabolism of toxins such as alcohol.

The body's peripheral timepieces have been likened to the instrument sections of an orchestra. The SCN is the conductor, coordinating the specific rhythms generated by these little clocks and synchronizing them according to light signals it receives from the outside world. But the peripheral clocks may step out and do their own thing—a phenomenon we may become aware of when we disrupt the symphony by traveling across time zones or working through the night.

What lies at the heart of every clock in the shop is a constellation of genes. Small variations in these clock genes may spell the difference between early birds happily up at dawn and those of more owlish bent, who struggle through the morning hours and hit their stride at midnight.

Louis Ptáček and his colleagues at the University of Utah were the first to reveal a direct genetic connection for extreme early birds. The team discovered that a large family of extreme larks living in Utah—patients with a disorder known as familial advanced sleep phase syndrome, who fell asleep as early as 7 P.M. and awoke as early as 2 A.M.—had a mutation in a central clock gene active in the SCN called *Per2*. Ptáček's team has since identified some sixty families with this gene. "These people were told they went to bed so early because they were depressed and antisocial," says Ptáček. Now it's clear they have a disorder related to their clock genes.

British scientists have shown that extreme larks and owls also tend to carry slightly different versions of the *Per3* clock gene. With remarkable consistency, very early risers had a longer variation of the gene than did late sleepers.

More moderate morning-evening preferences have also been correlated with such genetic variations. A team of scientists gave 410 subjects an owl/lark self-test to identify their preference for certain activities at certain times of day—time of rising, level of alertness on waking, favorite time for exercising and for doing mentally demanding tasks—to see where they might lie on the spectrum. The team also took blood samples from the subjects and compared the makeup of one of their clock genes. Those subjects with one variation of the gene showed a marked preference for eveningness, lagging as much as forty-five minutes behind larks in their preferred time for various activities.

As two prominent rhythm researchers have noted, "It seems that our parents—through their DNA—continue to influence our bedtimes."

Of course genes are not the whole story. Age matters, too. The transition from childhood to adolescence, especially, often sees dramatic shifts in avian tendencies. When Till Roenneberg studied the habits of twenty-five thousand people from the ages of eight to ninety, he found that children are typically early birds but start to become more owlish as they enter adolescence. The young child raring to go at 6 A.M. morphs into an adolescent who would rather not rise until noon—as anyone knows who has tried to haul a teenager out of bed for an early school starting time. On free days and weekends, adolescents will delay their sleep phase by almost three hours. This pattern persists until about age nineteen and a half for women, almost twenty-one for men. In fact, says Roenneberg, the peak in owlishness can be used as a biological marker for the end of adolescence. After this, the avian pendulum often swings back, and we return to a more lark-like pattern.

Light also has bearing: Research by Roenneberg suggests that many of us are owlish because we don't get the natural light needed to advance our clocks. People who stay outdoors thirty hours or more a week tend to go to sleep and wake up two hours earlier than those who stay out only ten hours a week. But even spending just an hour or two in natural light early in the day can advance your clock by as much as forty-five minutes. So, if you want to tip your body toward more larkishness, consider walking to work.

Young or old, lark or owl, we are not at our best on awakening. I recently took part in a psychological study that asked me to monitor my alertness over the course of a day. I carried a Palm Pilot at all times, and when it beeped, I responded to several questions, then took a quick test to measure my reaction time.

Early mornings were an embarrassment.

Even as a confirmed lark, I know I need time to sweep away the cobwebs of sleep inertia and meet the day in full alertness—time and a drug, specifically the potent stuff found in a mug of strong coffee.

I'm hopelessly addicted. Once, on a trip to a remote corner of northeastern China, I spent the night in an old army barracks that featured broken windows, a hole-in-the-floor toilet, and a mattress riddled with cigarette burns. Knowing that coffee would be scarce, I had brought with me ground beans and a French press for making my own brew. But boiled water was not to be had. I confess that for my morning rush I resorted to chewing the dry grounds.

The rich aroma, the rattling kettle: Just the ritual itself promises clarity.

Bach loved coffee. So did Balzac, Kant, Rousseau, Voltaire, who is said to have consumed dozens of cups a day—and my mother, who drank a relatively modest six. Two hundred years ago, Samuel Hahnemann noted that for coffee drinkers "sleepiness vanishes, and an artificial sprightliness, a wakefulness wrested from Nature takes its place." Today coffee beans are the most widely traded commodity after oil, and caffeine is the world's most commonly used psychostimulant drug. More than 80 percent of people consume it in one form or another, in coffee, tea, maté, cacao, kola. Members of the Achuar Jivaro tribe of the Amazonian regions of Ecuador and Peru wake up each morning by drinking an herbal tea made from the leaves of a South American holly, *Ilex guayusa*, which contains caffeine equal to about five cups of coffee. So strong is the concoction that men usually vomit up most of it to avoid the symptoms of overdose: headache, sweating, jitters.

To banish my own morning stupor, I depend on the buzz of the 300 to 400 milligrams of caffeine in two mugs of strong coffee, which I down at one sitting. New research suggests that taking your caffeine this way — in one big dose, Achuar Jivaro style — does not give you the most bang from your beaker. Charles Czeisler and his team at Harvard found that a single helping of caffeine may cause a quick peak in alertness, but it rapidly falls off. The most effective way to combat fatigue, improve cognitive function, and avoid the jitters is to take your coffee in smaller doses, two ounces every hour or so.

Just why caffeine has such a potent effect on the body has only lately come into focus. From the bloodstream, the chemical diffuses throughout the body's tissues and fluids, not stopping to collect in any particular organ but circulating evenly in blood — and in amniotic fluid and fetal tissue. It raises blood pressure slightly, dilates the bronchi of the lungs, and allows the body quicker access to fuels in the blood. In your kidneys, it increases the flow of urine; in your colon, it acts as a laxative. It even boosts metabolic rate a little, which slightly accelerates the burning of fat. Within fifteen to twenty minutes, 90 percent of the caffeine in your cup has left your stomach and intestines and begun to affect your brain.

The secret of caffeine's power as a stimulant is this: The drug binds tightly to the body's receptors for adenosine, a natural chemical important in sleep and wakefulness. As the cells in your body use up energy, they make adenosine as a byproduct; the harder they work, the more of the chemical they make. The adenosine attaches to receptors on cells everywhere in the body and quiets their activity. In this way it calms heartbeat, lowers blood pressure, decreases the release of stimulating neurotransmitters, and induces sleepiness. Caffeine enhances alertness by binding to adenosine receptors and inactivating them — thus preventing the chemical from exerting its quieting effects. So snug is the drug's fit with adenosine receptors that it's powerful even at low doses.

Caffeine works, then, not by exciting our nerve cells but by foiling the process by which they are calmed. Whether it actually perks up our brain function remains a topic of debate.

In 2005, a team of Austrian scientists used fMRI to watch caffeine's action in the brain. Prior to the test, a group of volunteers abstained from coffee for twelve hours. Then half the group drank two cups of coffee, and the other half drank a placebo. After twenty minutes, the subjects underwent fMRI scans while performing tasks involving memory and concentration. In all the participants, the brain regions involved in short-term or working memory fired up. But those who ingested caffeine also showed greater activity in the parts of the brain involved in attention and concentration (at least until about forty-five minutes into the experiment, when the activity tailed off). The researchers suspect that caffeine's effect on adenosine may be responsible for these regional boosts in neural activity.

There are naysayers, however. Roland Griffiths, a neuroscientist at Johns Hopkins University, suggests that the perceived mental benefits of caffeine many people experience from their morning coffee are an illusion: The coffee simply reverses the symptoms of withdrawal after overnight abstinence. Without coffee, says Griffiths, alertness would likely improve on its own an hour or two after waking.

Perhaps. But I can't wait. Illusion or not, I'm wedded to the quick chemical hit that jolts me out of morning muzziness and helps me make sense of the day.

2

MAKING SENSE

●

"Coffee?" I whisper to my sleeping husband. Though loath to startle him, I know my whisper is preferable to the shock of a bright light or the 70-decibel ring of his alarm. Morning comes to consciousness of sensory experiences, soft or jarring. Within a few seconds of waking, you can see stars, smell the dewy morning air, feel the light pressure of a sheet or the soft cotton of a shirt just slipped on, and recognize a partner's face or sleepy reply. Scent molecules waft up the nasal passages and latch on to receptors lodged in a tiny patch of tissue below and between your eyes. Nerve endings just beneath the surface of your skin detect the weight and texture of clothing, gentle as it is, and convert this mechanical energy into nerve impulses that your brain reads as touch, heavy or light, silky or scratchy. The sound of voice or buzzing alarm arrives on moving waves of air, which are eventually translated with exquisite efficiency into electrical signals interpreted as speech or birdsong or music. And even in the dim light of a dark bedroom, the forest of vision cells embedded in the retina captures the image of a face and flashes it to your brain.

At first it seems nothing could be simpler than this: the reliable registering of the world in one wide sweep through the discrete conduits of your five senses. Though it's a task that even the most powerful computer performs poorly, to you it seems as natural, as straightforward, as breathing. But as science has lately learned, there is nothing simple about it. A host of eye-opening new discoveries is complicating our view of perception, radically shifting it like the sudden twist of a kaleidoscope.

Take smell. Not long ago, your ability to smell—say, the overripe garbage or the fumes of your car warming up in the driveway—was considered a subpar skill of only minor importance, poorly understood and thought to engage only limited bits of your "lower" brain. Now smell is regarded as a highly sophisticated and sensitive system that can identify thousands of different odorants with some 350 distinct types of receptors, and analyze their dimensions in various regions of the brain to warn of danger or evaluate food. Our thresholds for detection of many odors are commonly in the parts-per-billion range, says Jay Gottfried, a neuroscientist at Northwestern University, "and we can readily discriminate between two different odors distinguished by only a single molecular component."

The odorants—complex organic molecules carried into the inner nose with inhaled air—meet the receptors in the mucus lining of the nose. Millions of olfactory nerve endings, each bearing dozens of identical receptors, poke into the mucus to interact with the world. The signal received by the receptor travels along the nerve by way of a long fiber, or axon, that threads through a tiny hole in the bone above it to the olfactory bulb region of the brain. In an amazing act of self-organization, the axons sort themselves so that the thousands of axons linked with neurons sporting the same receptors all converge in clumps at the same spot in the olfactory bulb. Each aroma sparks a constellation of these clumps, which the brain then interprets in various regions.

The character of a smell (fresh or foul? good or bad?) is sorted out in the orbitofrontal cortex, that all-important portion of the frontal lobe thought to be involved in decision-making, mood control, and drive. Its strength (how pungent?) is sometimes interpreted in the amygdala, the almond-shaped structure important in fear and other emotions—"but only when the smell is emotionally arousing," says Gottfried (for example, the reek of lion for a gazelle as opposed to the scent of a tree).

Identifying an odor, whether strong or faint, good or bad, enlists the regions of the brain involved in memory. A French study in 2005 showed that odor processing activates the memory regions in both hemispheres—probably, say the researchers, to help the mind gather relevant associations that assist in identifying the scent. As one researcher said, "We must first remember a smell before we can identify it."

Some odors may take you back along a deep groove of precise, personal memory. The aroma of bacon does for me, to waking summer mornings in childhood to the scent of thickly sliced farm bacon and smelts, perfect little fish my grandfather had caught fresh that morning from the dark waters of Lake Michigan and was lightly frying for his grandchildren's breakfast. For years anecdotal evidence has suggested that odors are especially powerful reminders of experience, an effect known as the Proust phenomenon, after the author's famous madeleine that summoned his childhood recollections. Scientists have found that olfactory stimuli do indeed evoke autobiographical memories more effectively than cues from other senses. And they fall away less rapidly than other sensory memories. This is even more astonishing when you consider that olfactory cells in your nasal epithelium survive for only a couple of months before they're replaced by new cells, which have to form new connections with cells deep in the brain.

What could account for the deep remembering of scents? Smell memories endure, according to the neurobiologist Linda Buck, because the olfactory cells that carry a receptor for a certain odor, whether they are new or old, always send their axons to the same spot in the brain.

The remarkable wiring of the olfactory system, it turns out, is also essential to taste.

There's nothing like that first sip of coffee. To get maximum pleasure from your mug, take a moment to savor the scent before sipping. The coffee vapors will pass from your mouth, around the soft palate, up into the nasal cavity, and thence to your olfactory bulb to whisper *java* to your brain.

Perhaps you thought your tongue responsible for the rich taste of coffee. But coffee's flavor—or any other flavor, for that matter—is mostly smell, about 75 percent in fact. Slurp a little Sumatra and your tongue will tell you only that it's bitter; that pleasant coffee taste, says Dana Small, is actually a pleasant odor pegged as taste because it's perceived to be coming from the mouth.

Small and her colleagues at Yale University discovered that the brain has a special sensory system devoted to odors delivered through the mouth. The team inserted small tubes into the noses of volunteers, one into the nostrils and one into the back of the throat. Then they introduced four odors into one tube or the other and scanned the subjects' brains using fMRI. The team found that for food-related scents, the two different delivery routes engaged different brain regions—suggesting that the brain possesses at least two distinct olfactory subsystems, says Small, "one specialized for sensing objects at a distance and one for sensing objects in the mouth." The latter is activated only when we breathe out through the nose between chewing or swallowing.

"A key fact about taste stimuli is that they elicit the most basic human emotions of pleasure (sweet) and disgust (bitter)," writes Gordon Shepherd, a neurobiologist at Yale. These are hard-wired in the brain stem from birth. By contrast, the responses to the odor component of taste "seem to be mostly learned," he notes, "which presumably accounts for the enormous diversity of flavours in the world's cuisines."

Very little was known about the real science of the mouth sense until quite recently. Now, test tubes, gene sequencing machines, and brain scanners are offering hints about what creates the full experience of flavor. The 25 percent wedge that is rooted in taste arises from receptor proteins that reside on taste cells within the taste buds of your tongue. Each of these receptors is dedicated to one of the five tastes: salty, sweet, sour, bitter, and umami. The latter taste quality (from the Japanese *umai*, for "good," and *mi*, for "taste") is responsible for the savory flavor of such foods as chicken broth, Parmesan cheese, mushrooms, and bacon.

Notwithstanding those tongue maps so ubiquitous in textbooks, the ones showing discrete areas sensitive to certain tastes—sweet on the tip, sour on the sides, and so on—cells responsive to the five basic taste qualities are scattered across the whole rolling landscape of the tongue. Though some taste cells are found on the pharynx, larynx, and epiglottis, most are located in the taste buds on the tongue.

Magnified, a taste bud looks like nothing so much as a little onion. Each bud possesses up to one hundred taste cells, which carry the receptors that do the real work of gustation: Chemicals in food slip through small holes in the buds, where they meet the receptors, which send their specific taste-quality message to the taste cortex of the brain. The brain then weds these taste sensations with information about fizziness and texture, the so-called mouth feel of food (which makes a crisp potato chip delicious and a soggy one unappetizing), and, in the case of hot chilies and other spicy foods, sensations of pain, to create the full-blown perception of the sweet, homey taste of banana bread or the savory flavor of squab infused with wine.

Even temperature enters the picture: Warming enhances the perception of sweetness and bitterness (another reason that hot coffee tastes so good). In fact, just changing the temperature of the tongue, cooling or heating it, will trigger an actual taste sensation in one out of every two people. In 2005, a team of researchers reported discovering the secret to the odd phenomenon known as thermal taste. When the tongue's receptors for sweet tastes are stimulated, a special channel opens. It turns out that heat, too, opens this channel, activating the taste receptors even when there's nothing to taste.

That all tasters are not created equal most of us know: Think of the sweet tooth that plagues some of us, and certain people's aversion to cilantro or anchovies. Think of George Bush Sr.'s well-known dislike of broccoli. Think of the flavor of olives, which to some is a divine mix of salty, sour, and bitter, but to others is similar to life at sea as Emerson described it, like being "suffocated with bilge, mephitis, and stewing oil." But only lately has it come to light just how wildly different a taste world each of us inhabits, especially where bitterness is concerned.

We humans possess a variety of some twenty-five bitter taste receptors, which are thought to have evolved to detect toxins in plants and foods. "Virtually every plant, edible or otherwise, contains toxins that can make us ill," says Paul Breslin of the Monell Chemical Senses Center. The picky eating habits of very young children, often directed against bitter-tasting fruits and vegetables, may be an evolutionary device to protect them from poisoning themselves when they're toddlers. Likewise, the nausea and aversion to certain foods experienced in pregnancy may have evolved to reduce a fetus's exposure to natural toxins. More women than men have a heightened reaction to bitter taste, though the sensitivity seems to vary over a woman's lifetime, rising at puberty and peaking during early pregnancy. After menopause, the sensitivity tails off, possibly because there's no longer a need to protect a developing child.

Scientists recently pinpointed small variations in these bitter receptor genes, which result in as many as two hundred slightly different forms of the receptors. Breslin has found, for instance, that people with a variant of one gene rate watercress, broccoli, mustard greens, and other such vegetables (which contain a compound toxic to the thyroid gland) as 60 percent more bitter than people with a different variant. So while you and I may share the template of two dozen or so genes for these bitter taste receptors, each of us carries his own distinctive versions, prompting either wrinkled nose or enthusiastic relish at the prospect of a plate of bitter greens.

Highly individual genes may also shape the way I make the morning's selection of tops and bottoms from my closet. Whether that vermilion shirt looks like a good match for those jade pants depends on the activity of genes that vary not only from person to person but from women to men—offering a possible explanation for heated spousal arguments over clothing and paint color. These genes were shaped in my primate ancestors' distant past.

I once had the good fortune to gaze into the eyes of one primate relative, a six-year-old chimp named Jack. For years I had heard about how similar we are to the chimpanzee in evolutionary terms, how much DNA we share, how close we are in anatomy and physiology. But nothing brought home the deep-down kinship like sitting face to face with the sensitive, intelligent, and *funny* Jack. There were differences, of course: Jack had a smaller head and bigger ears. His legs were short, his feet had thumbs, and he used his hands for walking. He didn't pray or sing nursery rhymes or tattle on his peers, at least not in a way that I could understand. But I found it startling and very moving to look into his eyes—darker, perhaps, but clearly akin to my own.

Jack loved nothing so much as the grapes and other small fruit snacks he received as rewards for his training. These he would balance on his lower lip thrust out as far as it would go, then slowly roll backward and flick into his mouth.

It's true that our eyes are the eyes of our predatory ancestors, insofar as they are set in front of the head in order to track prey with binocular vision. But human eyes, like chimp eyes, are also the eyes of the picky frugivores and leaf eaters that came before us, which may help to explain our idiosyncratic brand of color vision.

What allows me to see my clothes in a range of hues—scarlet, burgundy, turquoise, olive—is the interplay of three kinds of cone cells in the retina, each with a pigment especially responsive to light from a different portion of the visual spectrum: red, green, and blue. With this trichromatic, three-cone system, we humans can distinguish some 2.3 million gradations of color. So remarkably sensitive to the red/green segments of the spectrum is our tribe that we can tell the difference between the colors of light in this part of the spectrum with only a 1 percent difference in wavelength.

Our early mammalian ancestors saw the world in dichromatic vision, as most mammals do now, without the red part of the spectrum. Then, thirty or forty million years ago, the monkeys and apes of Africa—among them the primate ancestors of humans—experienced a mutation in a gene for a light-receptor protein that shifted its sensitivity from green light to red. It was a small change, but one that some scientists suspect gave our arboreal primate ancestors a clear advantage in the search for food, for picking out ripe fruits and tender red leaves against a background of green foliage. (This enhanced color vision may also have been useful for distinguishing other important objects from the surrounding foliage: brightly colored venomous snakes, for instance.)

New research hints at individual variations in this red-spectrum vision. When scientists recently analyzed a single gene that codes for a red-sensing protein in 236 people around the globe, they found 85 variants—about three times more variety than one might expect to see in other genes. This variation likely gives us each a unique view of hues.

Some percentage of women may experience even more distinctive color vision because they possess an extra red photopigment. If the visual cortex processes the additional input from this different class of red-sensitive cells, these women may be able to distinguish colors that look identical to the rest of us, allowing them to see a subtle color world forever unavailable to most of humanity.

It could be argued, then, that behind the simple act of everyday color perceptions—picking out a shirt from your closet, reading a traffic light, admiring a Rothko painting—lies a visual apparatus fine-tuned to locating red leaves and fruit, and an answer to the old philosophical dilemma, Is my red your red?

The answer is probably no. My experience of a tomato likely differs from yours, both its exuberant red hue and its tartly acidic tang. As the great psychologist William James said, the mind works on the data it receives "very much as a sculptor works on his block of stone. Other sculptors, other statues from the same stone!"

If we owe our brilliant color vision to our ape ancestors, we may be obliged to another creature for our fine sense of hearing. As you get dressed or make lunch or tidy up before you leave for the office, you listen with one ear to the morning news while the other is tuned to a volley of conversation from your spouse settling plans for the evening, or to your children searching for school books, or to the annoying, persistent barking of your dog from the far end of the yard. How in the world do your ears pick up the subtle vibrations of sounds, faint and furious, and sort the cacophony into intelligible, sensible parcels?

Detecting the source of sound and interpreting it—a Bach sonata or your teenage daughter's plea for a matching sock—may seem like an easy feat, but it's exquisitely complex. When we hear someone call our name and turn toward the sound, we are relying on the brain's ability to calculate direction from the interaural time difference, or ITD—the difference in the time it takes for the sound to reach our two ears. "Incredibly, we can detect ITDs of only a few microseconds, allowing us to distinguish between sounds separated by only a few degrees in space," writes the neurobiologist George Pollak.

Our refined ability to parse sounds in time and locate them spatially we may owe to the dinosaurs, which forced our earliest mammalian ancestors to retreat to a nighttime niche. For millions of years, our shrew-like forebears carried out their lives under cover of darkness, where sound held sway over sight. Over the eons, they evolved a highly sophisticated auditory system incorporating the time dimension. Now our ears can perceive sounds lasting only a fraction of a second in their correct order and locate them in space.

Sounds arrive at the ear as waves, which your eardrum, or tympanic membrane, converts into energy that rattles the three delicate little bones of the middle ear. This causes pressure changes within the cochlea, the coiled, fluid-filled tube at the heart of the ear, which in turn translates the energy into chemical and nerve signals that are sent to the brain.

The cochlea is no passive spiral cavity, as once believed, but, in neuroscientist Jim Hudspeth's words, "a three-dimensional inertial-guidance system, an acoustical amplifier, and a frequency analyzer compacted into the volume of a child's marble." Our ability to hear depends on "hair" cells, arranged in the cochlea in a zigzag pattern. These cells are relatively scarce—only sixteen thousand to an ear—fewer cells, Hudspeth notes, than you have on a hangnail or a flake of dry skin—which accounts for the vulnerability of our acoustical system. Hair cells damaged by infection, drugs, aging, or excessive exposure to Deep Purple are lost for good.

If I put a small microphone to my sleeping husband's ear, I might well hear his hair cells hard at work. In a quiet environment, the hair cells in most normal human ears are turned up to amplify softer sounds—turned up so far that they themselves generate faint but constant tones of sound, like the feedback noise from an electronic amplifier. In a loud environment, a thunderstorm or rock concert, the hair cells adjust, turning down their amplifiers. It is thanks to these mini-amps that we can follow ten to twenty distinct sounds per second, distinguish pitch, and hear noises that last only a few thousandths of a second.

We rarely notice the sound made by our hair cells because the brain filters it out. Likewise, when we speak or sing or vocalize in any way, the brain halts the firing of our auditory neurons so that we won't be swamped by our own song. So, too, the brain allows us to suppress a whitewater of auditory stimuli—the buzzing, banging, humming, thumping background noise of our typical morning routine—so that we may hear only what interests us; the rest fades into a kind of muted roar that we hear with just "one" ear at first, then with no ear at all.

This is one example of desensitization, the same phenomenon that makes the aroma of bacon or reeking garbage fade from perception, that helps our eyes adjust to bright light, that allows us to forget the rub and weight of clothes on skin, and that attenuates the nervous jolt initially provided by coffee. Desensitization can take place over seconds (light), minutes (smells), or days (caffeine).

At any given moment, we tune in to what's important in our world by turning off stimuli. We also fill in what we are missing. Think of conversing over the ambient noise of the morning radio. Often you hear only part of the conversation (the rest being masked with radio chatter), but nevertheless grasp the gist of it by filtering out the irrelevant noise and filling in the missing sounds.

Something similar happens in the brain when you sing a song inside your head. In 2005, scientists scanned the brains of subjects while they listened to the soundtracks of familiar songs (the Rolling Stones's

"Satisfaction," for example, and the theme from *The Pink Panther*) in which silent gaps had been inserted. The auditory cortex of the subjects continued to show the same pattern of activity even when there was silence on the soundtrack and the subjects were just "singing" in their heads. The ear wasn't hearing the song, but the brain was.

Sensing isn't what we thought it was. It's a far more sophisticated endeavor shaped by our genetic makeup, our creative powers of filtering and filling in—and, quite possibly, some significant crosstalk between the senses. I've been discussing the morning as if we sensed its elements separately, one facet at a time, but in fact the brain is always uniting the different qualities of individual objects so that we don't associate the color of one thing with the movement of another—so, for instance, we can see a cat as black, shaped like a cat, and meowing, and a yellow dog as dog-shaped, yellow, and howling. Scientists are still searching for the "glue" that binds together different sensory aspects; some theorize it may be the synchronous firing of the neurons from different areas of the brain involved in perception.

What if we did process only one sense at a time, if you took in only the sight of your child's face and not the timbre of her voice, or if you could only smell your morning juice but not see it? Would it taste the same?

Probably not. What you see changes what you taste. When a French scientist, Gil Morrot, gave a panel of fifty-four tasters a white wine artificially colored red, the group—experts and nonexperts alike—described its odor and taste as that of a red wine.

Likewise, what you see affects what you hear and feel. In one study, researchers placed monkeys in the center of a semicircle of speakers and trained them to look in different directions while they listened. Then the team monitored the signals arriving in the part of the monkeys' brain that transmits information from the ears to the auditory cortex. To their surprise, the cells in this region fired at a different rate depending on where the monkeys' eyes were directed.

Similarly, scientists have found that when people look at a spot on their body where they're being touched, their brains show greater activity in their somatosensory cortex—the touch region of the brain—than if they don't see the touch. The reverse is true as well. Giving people both a touch and a visual stimulus at the same time and on the same side of the body enhances activity in the visual cortex.

So vision is never just seeing, touch is never just touch. We spot objects more easily if we hear a relevant sound simultaneously. When we see a banana or a crimson shirt, we're also feeling it with our mind's "hand."

This crosstalk occurs in sensory memories, too. Jay Gottfried and his colleagues have found that a memory cue in one sense reactivates other sensory memories. Most of us know this from experience—the smell of coconut oil elicits the intertwinkling facets of that white stretch of beach and light slap of waves; the scent of smelts evokes my grandfather's kitchen, his cigar smoke, and his smile with its flashing gold tooth.

Our senses, then, are hardly the simple, discrete instruments we once imagined, but picky, idiosyncratic tools that interact with subtlety and speed to grant us out of mere electrical disturbance our own distinctive view of . . . of what?

Whatever we're paying attention to at a given moment—say, the traffic during our morning drive to work.

3

WIT

●

OU'RE OUT THE DOOR and on your way, cruising along at
forty or fifty miles per hour, your mind—well, your mind is
not entirely on the highway along which your two-ton steel
bullet is hurtling. You may think you're absorbing the scene in all its
detail, four-lane road, swerving Subaru, white morning glare, but the
impression you have of seeing everything is an illusion. Though your
senses are taking in some ten million bits of information a second,
you're consciously processing only about seven to forty bits. Even
fewer if your thoughts are really elsewhere—on your upcoming meeting, say, or replaying that morning family spat or perhaps trying to
settle it now by cell phone as you drive.

"We actually only see those aspects which we are currently visually
'manipulating,'" says the psychologist J. Kevin O'Regan, and we visually manipulate only the things we're paying close attention to.

This came home to me one cold winter morning a few years ago
when I stood with my ten-year-old daughter, Zoë, watching the waking of whooper swans from the bank of a calderic lake in Hokkaido,
Japan. The lake was ringed with blue volcanic hills and fed by hot
springs; on the shore near us was an outdoor Japanese bath. My eyes
were all for the swans—pale white mounds of feathers, heads tucked
beneath wings—so as to note the pattern and behavior of their waking. One by one the necks of the birds before us unfurled and their
heads popped free of their wings. But what was that low, furtive shape

slinking along on the ice behind them? A dog? A fox? So focused was I on the furry form that I failed to notice another dark figure not ten yards from us, a man walking buck-naked to the baths.

Zoë saw him, all right.

My failure to spot the obvious is an example of "inattentional blindness." When the brain is attending to its surroundings, it is aroused, ensuring full awareness and efficient activity. But when it's distracted, it's capable of missing what's most obvious. This is the phenomenon demonstrated in those "man in the gorilla suit" studies showing that people told to engage in a simple task, such as counting the number of shots taken during a basketball game, will completely miss the man running across the court dressed in a gorilla suit. It also underlies what my family calls refrigerator vision, when you miss what you are looking for—the obvious jar of mayonnaise or leftover lasagna—because someone is asking for the ketchup.

Francis Crick and Christof Koch suggest that our ability to take conscious notice of an event has to do with the way attention influences the coalitions of neurons responding to various sensory stimuli —the deer by the highway, the distant whine of a siren, the naked bather. These coalitions vary in size and character, Crick and Koch suggest; they form, grow, compete with one another, disappear, or endure in fluid response to changing situations. Only those that are sustained make their way into consciousness as a registered perception. Attention, the theory goes, somehow serves to determine which rival coalitions win the competition. Perhaps attention jacks up the neural activity in one coalition by firing a certain way, making the stimulus that activated that coalition seem bigger and brighter than those activated by competing stimuli. In this view, attention not only points a finger at a sensory experience, it *makes* the experience.

Even when we think we're fully attending, we may miss critical particulars. Imagine this challenge: As a rapid stream of numbers flash by, pick out the two letters among them that appear randomly for only a tenth of a second. How would you do? Chances are you'd pick out the first target letter, but if the second appeared within a half second of the first, you'd fail to see it. This is because of a weird neural bottleneck—an attentional "blink" that prevents you from consciously attending to multiple visual events so close together in time.

Which begs the question: What happens when we try to attend to two things at once?

• • •

For years my mother shuttled my profoundly retarded sister back and forth to her special school—a tediously long, twice-daily commute. While she navigated the Virginia highways, she would drink a cup of coffee and memorize poetry from a book propped up on her dashboard. For her, such juggling of Sumatra, steering, and Wallace Stevens was a kind of mental necessity to keep her mind nimble during the numbingly long drives. But most of us multitask out of an obsession with making the most of our time: We listen to the radio while reading the newspaper; we pay bills while chatting on the phone and type text messages while sitting in meetings.

How efficient is this? Are we doing both jobs justice? Are we saving time?

"To do two things at once is to do neither," wrote Publilius Syrus in A.D. 100. Evidence is building to support the Latin poet. Despite the massive parallel-processing capabilities of its one hundred billion neurons, the brain just isn't built for two-timing. In attempting two tasks at once, it can be foiled by even the simplest jobs.

Take the matter of gauging how much time you have before the signal light ahead switches from yellow to red. To brake or sail through? The answer depends in part on your interval timer, another clock possessed by your brain. This one is expert at gauging the passage of time from seconds to minutes and hours. We tend to have pretty good interval sense when we're paying full attention—accurate to within 15 percent—and use it to make decisions and judgments in all sorts of everyday situations: running to catch a bus or a baseball, singing along with Regina Spektor, keying in a number on a cell phone while glancing up periodically to check the road. But as science has discovered, this internal timer suffers mightily from distraction.

Just how our brain calculates time intervals has been one of the most elusive concepts in neurobiology. Unlike seeing or hearing or smelling, interval timing has no dedicated sensors, as Richard Ivry, a cognitive neuroscientist at the University of California, Berkeley, points out; nonetheless, it's as "perceptually salient as the color of an apple or the timbre of a tuba," he says, and we need it for driving, walking, conversing, playing music, participating in sports, and a million other daily activities. For years scientists thought that the interval timer was located in some central hourglass area of the brain, inspired, perhaps, by the discovery of the circadian master clock in the SCN. But new research indicates that the brain may judge intervals through the activity of a network of neurons widely scattered among many different brain structures—and that different time intervals may be processed by different neuronal networks. Ivry's work suggests that the cerebellum, the part of the brain that coordinates movement, plays a role in the timing of tasks in the range of milliseconds. For longer intervals such as timing a yellow-to-red stoplight, the brain most likely uses a more distributed system, says Ivry, involving structures active in working memory, such as the prefrontal cortex and basal ganglia.

Temperature may toy with this clock, disrupting our ability to accurately time intervals that span more than a second. A doctor discovered this when his wife was ill with a high fever. He rushed out to the drugstore to get her some medication; when he returned twenty minutes later, she was upset with him for taking so long, claiming that he had been gone for hours. Intrigued by her misperception of passing time, the good doctor asked her to estimate a minute by counting to sixty at one number per second. Her estimate turned out to be thirty seconds long. As her temperature fell, her performance improved.

However, nothing flummoxes our interval timers like distraction. When participants in one study were asked to assess intervals of fifteen to sixty seconds while they performed concurrent real-life tasks, their accuracy plummeted. When you're engaged in one thing, time expands. When you're dual-tasking, it contracts; the brain misses the metaphorical "tick" of some number of pulses, so time seems shorter. It's simple: Accurate judgment of time requires attention to its passage—of critical import in matters of traffic. This is one reason why driving and chatting on a cell phone is not a good idea. There are others.

I'm no good at multitasking. When I talk on the phone, I can't hear my husband's verbal prompts or read his written ones. I can't drive and change a CD, much less memorize poetry. Not long ago, at a psychology lab at the University of Virginia, I had my inadequacies officially documented. But I also learned that I'm not alone in my "disability": Most people overestimate their capacity to attend to two things at once—especially when they're driving—with consequences that range from annoying to catastrophic.

"What do you get when you cross an owl with a goat?" This was the riddle posed to me the day before by Bryan, the first-grader I tutor in reading, and his teacher. Only later, in the small basement room of the Cognitive Aging Laboratory, was the answer percolating—when my mind should have been on the task at hand. My mission: to write down all the words I could think of in one minute that begin with the letters *f, a,* and *s.* I got started with a few familiar verbs and objects—animals, furniture, fruits—but then stopped. *F? S?* My mind suddenly locked up. Words starting with *a*? I couldn't think of a single one. Then "agoraphobic" popped out of the box, and "soporific" and "flagrant" and "felicitous." I was keenly aware that these long Latinate adjectives were wasteful luxuries; I should stick with the Anglo-Saxon, the monosyllabic—the sip, sap, soap; the flea, fly, and feel. My back muscles tightened, my hands grew clammy. I tried to think of the *fl* consonant blends I had supplied to Bryan—flip, flop, flap. Then the riddle seeped into consciousness to steal my attention.

Hootenanny.

That Bryan could appreciate such wordplay struck me as a small miracle. One of the many children who for one reason or another are deemed "at risk," Bryan had come to town only a few months earlier with his mother and older sister and virtually nothing else but the clothes on his back and a particular shade of sweetness that made teachers, librarians, and custodians silently mouth at me as we passed in the school halls, "I love that kid." When Bryan arrived, he had struggled with the very basics of language, with rhymes and phonemes, those sound packages that make up words. Just a few days earlier he had stumbled on a word that stumped him. "Wish?" he inquired. "What's a wish?"

The question had stunned me. Think of the myriad folk- and fairy tales most of us encounter in childhood, of wishes granted and often squandered by luckless recipients: Cinderella, the Frog King, the Seven Ravens, and of course the Grimm brothers' Three Wishes, a story I remember first reading when I was Bryan's age. After a minor bout of misbehavior and subsequent punishment, I had sought refuge with my book of fairy tales in my parents' clothes closet, dimly lit by a small circular window and redolent of shoe polish, mothballs, and my father's after-shave. I remember especially the image of the wished-for sausages stuck on the nose of the poor wife, with only a single wish to go.

Somehow, in the midst of all his want, Bryan had missed the concept of wish. I had asked him to tell me his big three. "I *wish* for a Popsicle. I *wish* for new sneakers. I *wish* for a remote-control car." He paused for a minute, then grinned at me and sang out, "WISH, FISH, DISH, PISH, MISH!" That Bryan could come so far in such a short time, to get the puns and verbal jousting, was humbling testimony to his powers of focus and concentration.

My minute was up. I flashed a smile of chagrin at the twenty-something graduate student administering the test. He seemed to have little sympathy. There it was in raw data: The split attention of this writer yielded an abysmally poor score on, of all things, verbal fluency.

The fluency task was the first of a dozen or so cognitive tests I would take over the next several hours as a participant in a study of the brain and its workings by Tim Salthouse, the director of the laboratory. Though the setting was artificial and the tasks contrived, I could see how these tests were aimed at parsing the things our brains do on a daily basis. They are windows on how we think—specifically, how our brain accomplishes its repertoire of organizing activities, called executive functions: focusing attention, concentrating on what's relevant and ignoring what's not; making split-second decisions, often based on conflicting information; shifting mental goals and rules in the face of new demands; doing two mental jobs at once.

Among the trials were classic tests of dual-tasking: steering the wheel of a driving simulator to keep a tracking ball between two wiggly, swerving lines while counting backward by threes from 862; and the Stroop test, a list of color names presented in mismatched ink (the word "blue," for instance, printed in red). The subject is expected to quickly reel off the names of the printed colors, not the words.

My performance on both Stroop and simulator was pathetic— though, it turns out, not all that much worse than the average. Teens who play video games are sometimes pretty good at the simulator task, but the Stroop test often stumps even the young. Because reading is more automatic than recognizing and naming colors, speed in this test requires focusing attention only on the color of the typeface and simultaneously inhibiting the desire to read its verbal content. Try it yourself; it takes longer to say "red" to the word "blue" printed in red than to say the word "red" printed in red, because two mental processes are in conflict. (Rumor has it that the CIA used the Stroop test in the 1950s to ferret out Russian spies. The names of the colors were written in Russian; if test participants were slowed by the written words, it was a sign they knew the language and might be spies.)

People's ability to dual-task is a good deal less impressive than they think, in part because of the limitations of working memory. If you can recall the start of this sentence while you read the end, you can thank your working memory. Also known as short-term or scratch-pad memory, it's what allows you to keep in mind several facts or thoughts (most people can hold between five and nine) and manipulate them for a brief time, a few seconds or so, while solving a problem or performing a task: holding in mind a phone number, for instance, or remembering where you are in the subtraction part of your simulator task, or recalling how long it normally takes for a traffic light to turn while deciding to brake or accelerate.

When you try to carry on a cell phone conversation and stay on the road, tuned to traffic, you strain your working memory as well as other executive functions, such as the ability to shift goals in your mind, activate new rules, and refocus your attention.

To quantify just how efficiently the brain toggles between two mental jobs, David E. Meyer and his colleagues at the University of Michigan asked a group of participants to perform two tests of dual-tasking. In the first, they were asked to switch repeatedly between a pair of tasks focused on geometric shapes, judging one perceptual characteristic (shape, for instance) and another (color, size, or number); in the second, between two tasks involving different kinds of arithmetic problems (say, switching back and forth between multiplication and division). In both tests, the participants took more time to complete the tasks simultaneously than they would have if they had performed the tasks one after the other. "Sometimes they showed an increase in total task completion time of 50 percent or more," says Meyer. That's because the brain takes time to shift its mental rules and goals, he explains, to leap from "I'm doing this, which requires these rules, to I'm doing that, which requires those rules" — several tenths of a second, in fact, which adds up if you're toggling a lot.

When you're on the cell phone and the vehicle you're driving is moving along at eighty-eight feet per second, those lost moments may mean the difference between life and death. According to a 2006 study by the National Highway Traffic Safety Administration, some 80 percent of crashes and 65 percent of near crashes have involved some form of driver inattention less than three seconds before the event. Talking on a cell phone increased the risk of a crash or near crash by 1.3 times; dialing a cell phone tripled the risk.

• • •

It's 10 A.M. You've reached the office safely and swallowed a second cup of coffee while returning phone calls and e-mail. In another hour or so you're expected to give a presentation at a meeting. Now you plunge into the required reading, all fine focus and concentration. If we could eye the workings of your gray matter as you scan those pages of dense text, what would we see? What's going on beneath the snug carapace of your skull while you engage in keen-witted reading (or, for that matter, its nemesis, distraction)? Until lately, the brain and all that it lodged—the ability to think, feel, act, imagine, reason, remember—were one dark riddle. But in the past decade, science has opened stunning new windows that allow a look inside to see in detail a brain going about its business in real time.

One late morning in a laboratory at the Yale School of Medicine, two neurologists, Sally and Bennett Shaywitz, are doing just that: watching the activity inside the brain of an engaging eleven-year-old boy named Keith. Through a plate-glass window, I can see Keith lying on his back with his head in the circular scope of an MRI scanner. He's reading a series of paired cues through a periscope—a word and picture simultaneously flashed on a screen ("fox" and the image of a box, "cow" and a bow)—then quickly pressing the yes or no button on a button box to indicate whether or not the two rhyme.

The Shaywitzes are probing the circuits in the brain involved in reading. Right now they are hovering over two computer screens, one displaying the set of constantly changing reading cues supplied to Keith; the other showing a monochrome image of his brain in cross section. The scanning results in structural images, revealing the finest details of brain anatomy, and also functional images, showing the location of the brain's activity.

An MRI scanner is safe and noninvasive, requiring no radiation or injections. A massive and powerful circular magnet, it looks, as Keith says, like a spaceship or a doughnut filled with milk. MRI scans can provide a detailed anatomical picture of the brain with a resolution of less than half a millimeter, explains Sally Shaywitz, fine enough to detect an artery the diameter of a hair deep in the center of the brain.

As Keith reads his series of cues, computers are also compiling data on the neurons activated in his brain. Functional MRI reveals brain regions active during the performance of specific tasks by monitoring the changes in oxygen and blood flow that accompany neural activity. The harder a particular region of the brain works, the more blood-borne oxygenated hemoglobin moves into the area. A "blush" of this hemoglobin registers on the MRI scan as a slight rise in signal strength. In this way, the scan produces pictures of the cell circuits that fire as we engage in a particular mental activity. After the data are compiled, the result is a series of color photographs showing different brain areas "lit up" in a rainbow of hues, a kind of moment-to-moment map of neural activity.

Neuroimaging is not without its critics, in part because of the time scale of the technology. Functional MRIs take pictures on a scale of seconds; neural firing occurs on a scale of milliseconds. Furthermore, the activity that shows up in an fMRI is not necessarily causal. The scans show which regions are active during cognitive tasks but not necessarily which are essential to that task.

Still, says Sally Shaywitz, "functional neuroimaging has revolutionized the way we see the working brain. It can take a hidden function—and dysfunction—and make it visible." Such studies have put to rest the myth that we use only a small portion of our gray mass, the proverbial 10 percent. In fact, most of the neuronal nooks and crannies of the whole buzzing, blooming thing are fired up in the course of a day—though not all at once. Different groves of neurons erupt into activity at different times, with different tasks. Scans have captured the brain in action, navigating, calculating, grasping language, recognizing faces and places, perceiving time, reading verbs.

Imaging studies by the Shaywitzes and others have pointed to very specific neural areas that grow active in reading. Among these are a phonological region in the rear brain, just above and behind the ear, used by novice readers such as my student Bryan to sound out words phoneme by phoneme; and behind this, a so-called word-form area in the occipitotemporal region of the rear brain, which allows the expert reader to recognize whole words extremely rapidly, in less than 150 milliseconds. As learning readers move from novice to pro, they shift from using their phonological region to relying primarily on their expert word-form area.

It is this skilled circuit that flashes with activity as you pore over your work. The occipitotemporal region also grows active in the brains of car experts as they look at various makes and models of classic cars, and in ornithologists when distinguishing between different species of warblers. In fact, this rear-brain area may be important for expertise of all sorts, says Bennett Shaywitz. "It seems to be good for learning expert tasks, for getting better and better at something."

• • •

You hope your rear-brain expertise circuits have kicked in during your morning work and that you are now primed for your presentation. Your meeting has begun, and you're feeling sharp and on your toes. Late morning is a peak time for certain kinds of mental activity, according to some chronobiologists. Studies show that alertness and memory, the ability to think clearly and to learn, can vary by between 15 and 30 percent over the course of a day. Most of us are sharpest some two and a half to four hours after waking. For early risers, then, concentration tends to peak between 10 A.M. and noontime, along with logical reasoning and the ability to solve complex problems.

Much depends on age, however. For teens and young adults, morning may feel a far cry from Rilke's "bright new page." Mary Carskadon, a chronobiologist at Brown University, has documented in longitudinal studies the physiological change in the body's biological clock during the adolescent years. Older teens shift toward a more owlish, or phase-delayed, pattern, secreting the hormone melatonin later in the evening and delaying their bedtimes. Yet they're forced to wake up early for the start of school. "Requiring older adolescents to attend school and attempt to take part in intellectually meaningful endeavors in the early morning may be biologically inappropriate," says Carskadon. Not only are these teenagers sleep deprived, "but they're being asked to be awake when the circadian system is in its nocturnal mode. The students may be in school, but their brains are at home on their pillows."

The relationship between the circadian rhythms of the body and mental performance is subtle and still a matter of debate. How well you do at a given mental task may be affected by a host of variables —boredom, distraction, stress, how confident you feel, how much sleep you got the night before, what you ate for breakfast, whether you had caffeine, your posture, the ambient temperature, air quality, noise, lighting, and other "masking" factors that have little to do with your circadian rhythms. "Time-of-day effects are intriguing but controversial," says Tim Salthouse, because they're difficult to isolate and replicate in scientific studies.

Still, there's evidence to suggest that the daily ups and downs of body temperature influence mental performance, with predictable peaks and troughs. Some studies have shown that the function of neurons is affected by brain temperature: Higher temperatures may result in faster transmission of impulses between neurons. Scientists at the University of Pittsburgh tested young adults over a thirty-six-hour period, taking their temperature every minute and measuring their performance every hour on various tasks of speed, accuracy, reasoning, and dexterity. The team found a significant time-of-day variation, with a nocturnal trough in performance close to the lowest body-temperature readings. On the flip side, researchers at Harvard reported finding a correlation between higher body temperatures and peak performance in alertness, visual attention, memory, and reaction time.

Two mental functions may be particularly susceptible to subtle circadian variations, according to Lynn Hasher of the University of Toronto and her colleague Cynthia May of the College of Charleston: decision-making and "inhibition"—the capacity to suppress distracting, irrelevant, or off-task information (such as the verbal content of those color words in the Stroop test). At "off-peak" times, people are more likely to have trouble suppressing distractions and to fall back on accessible, familiar decision-making routes rather than ones that demand analysis and evaluation. Work by Hasher and May suggests that these more subtle circadian effects vary with age. Young adults "are clearly bothered by distraction in the morning," say the researchers, "but, later in the afternoon, it is as if the distraction were invisible to them. The data for older adults show the opposite pattern."

Because inhibition is particularly difficult at one's "off" times, May recommends that people restrict to their peak hours those tasks requiring "focused attention (e.g., reading complex instructions), retrieval of exact information (e.g., recalling medication dosages), or careful control over responses (e.g., driving in heavy traffic)," or at least try to complete them in a setting with minimal distractions. On the other hand, as May points out, there may be some advantages to low inhibition. In tasks requiring creative problem-solving, less inhibition may allow people to consider more imaginative solutions.

Memory, too, may fluctuate with time of day. According to Hasher's work, older adults tend to experience what she calls "a substantial increase in forgetting across the day": Mornings, they forget an average of five facts; afternoons, about fourteen. The reverse is true for young adults.

In the past few years, scientists have begun to trace the role of circadian rhythms in learning and memory right down to the level of molecules, all with the help of a giant snail, *Aplysia californica*. If perchance you tried to master the material you needed for your meeting by staying up until the wee hours of the morning, performed well in the presentation, but then later found that your memory was vague on what you had learned, you're in good gastropod company.

Why *Aplysia*? "It may not be a beautiful animal," says Eric Kandel, "but it's extremely intelligent and accomplished, with the largest nerve cells in the animal kingdom." A Nobel Prize–winning neurobiologist at Columbia University, Kandel has seen firsthand what the humble sea snail can tell us about what's happening in the brain when we glean new knowledge from our reading or absorb a lesson from a colleague or teacher.

"We humans are what we are because of what we learn and re-member," says Kandel, "and in a way it's mind-boggling that we know what we know about what changes in the brain when we learn something new—how our minds are different at the beginning of a learning experience than they are at the end—because of studies of *Aplysia*."

Kandel has been fascinated by the puzzle of learning and memory for more than a half century. Born in Vienna in 1929, he grew up in a morass of barbaric human behavior. He was taunted for being a Jew, watched his father seized by police, and at the age of nine witnessed the horror of *Kristallnacht,* which he remembers, he says, with a kind of flashbulb memory, "almost as if it were yesterday." In 1939 he and his family fled Vienna. Kandel spent the rest of his life asking ques-tions about the nature of the mind: why people behave the way they do, how they hold on to the memories that shape them, and above all, how they learn. He believed that insight into the nature of our own being could arise through the study of lower organisms.

Indeed, from the nerve language of *Aplysia* Kandel garnered one of the great secrets of the human mind: Learning results from changes in the strength of the synapses, or junctures, between two precisely interconnected brain cells. In creating short-term memory, the brain strengthens already existing synaptic connections by modifying pre-existing proteins. In making long-term memories, it makes new pro-teins and grows new synaptic connections.

Though the process may be a good deal more complicated in hu-mans than in marine snails, says Kandel, it appears to involve a similar set of mechanisms. At a much simplified level, it may go something like this: At any moment, your brain is alive with firings. A single neuron receives a stimulus and fires here, triggering another neuron to fire there. Much of the time nothing comes of the activity. The chemical message one neuron sends to its neighbor may be too weak or sporadic to set the neighbor alight and form a network. But when the mind is focused and attending, as it is during learning, that single neuron may send more frequent, stronger messages to its neighbor. The synapse on the neighboring neuron is then chemically altered by the exchange. If the first cell fires again, even weakly, it may trigger a synchronous response in the now more receptive second cell. This leaves both cells aroused and ready to fire again in the same pattern.

The upshot of all this may be merely a transient notion, flashed briefly across the mind, a memory trace that lasts barely a few seconds before it passes into oblivion. But if the stimulus is repeated, and the neurons continue to fire in synchrony, the synapses between them will strengthen. Eventually they are bonded, so that when one fires, the other does too. This process of bonding between neurons, known as synaptic plasticity, may be what underlies both learning and memory.

Once the process has taken place, Kandel's theory goes, signals blaze more easily along the pathway between the neurons, and the same signals produce larger responses. If the activity is repeated—the remembering of a word or concept or skill—the linking and lock-step firing continues and spreads to other neurons, forming a network of well-bound neurons, which fire together in the same pattern each time they are activated. The process draws together neurons involved in an event or idea. Hence the expression "Cells that fire together, wire together." With each repetition of a skill or activity, with each additional firing of the circuit, the synapses become more efficient and the learning more permanent.

"Practice makes perfect," says Kandel, "even in snails."

Aplysia is in the learning limelight again, this time for what the gastropod tells us about circadian effects on learning and memory. In 2005, researchers at the University of Houston reported finding that the snail suffers forgetfulness when it pulls all-nighters. Like us, *Aplysia* is a diurnal creature that prefers daytime life. To probe the influence of circadian rhythms on its patterns of learning, the team examined its ability to absorb and remember lessons about noxious substances and inedible food. The study showed that *Aplysia* forms short-term memory of lessons equally well during day and night, but builds long-term memories only when it's trained during the day. At night, say the researchers, the biological clock seems to shut off the proteins involved in forming long-term memories—a lesson perhaps worth remembering.

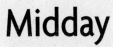

Midday

Think in the morning.

Act in the noon.

WILLIAM BLAKE

4

THE TEETH OF NOON

•

YOUR MEETING HAS EATEN into the lunch hour. Breakfast was skimpy, and now, five hours later, it's getting harder to attend to the matter at hand as your thoughts turn inexorably to the sushi buffet at your favorite Japanese restaurant or the hearty ham sandwich stashed in your lunchbox. The fifteenth-century Venetian surgeon Alessandro Benedetti asserted that nature had relegated the stomach to a site distant from the brain, fencing it off with the diaphragm, "in order not to disturb the rational part of the mind with its importunity." Nature, it would seem, failed in its mission.

What does the mind look like while pondering unagi or honey-baked ham? Where does the craving originate, in the belly or the brain? One would assume clues could be found in people who think incessantly of food. Not long ago, two Swiss researchers, neuropsychologist Marianne Regard and neurologist Theodor Landis, conducted a brain imaging study of such people, a group of patients afflicted with a benign eating disorder the scientists termed gourmand syndrome.

The syndrome was first identified by the team in two patients who developed obsessions with food after suffering stroke damage to their right frontal lobes. Before their illnesses, both patients had been average eaters with no particular food preferences. After his stroke, one patient could think of nothing but tasty food served in a fine restaurant. "It is time for a real hearty dinner," he wrote in his hospital diary, "a good sausage with hash browns or some spaghetti Bolognese, or risotto and a breaded cutlet, nicely decorated, or a scallop of game in cream sauce with 'Spätzle' (a swiss and southern german specialty). Always just eat and drink!" The second patient experienced similar cravings and a yen for food shopping, cooking, and selecting restaurants. He also got a thrill out of recounting special meals: "The creamy pastry slips from the foil, like a mermaid," he wrote. "I take a bite. From now on, it will be more difficult to put me under stress."

To follow up on their observations, the Swiss researchers scanned the brains of thirty-six other passionate eaters and discovered that thirty-four of them had lesions in their right frontal lobes. The scientists were quick to say that their findings do not point to this right-hand corner of the brain as a food-contemplation area, but rather an area possibly involved in impulse control and obsessions of all sorts.

Still, I found the mania curiously familiar and wondered whether it's possible to experience this sort of frontal-lobe activity in varying degrees. I admit to more than a smidgeon of the syndrome myself, a tendency to think too often of food and to remember meals in inordinate detail: the artichokes stuffed with shrimp served on a deck in Fresno, the fried catfish dished up with collard greens in a juke joint in the Delta, a root beer float gratefully inhaled on a lakeshore at my first sleep-away camp. (Letters home from this camp were a litany of complaints about the food, save one: "I may seem happy in this letter but it's only because we had French toast this morning.")

My husband once ate a meal in Julia Child's kitchen and can recall only that he was served "some sort of chicken"—a failure of culinary consciousness I simply can't fathom.

The sort of intense preoccupation with food suffered (or enjoyed) by people with gourmand syndrome may lie at the extreme end of the scale, but all of us fixate on food when we haven't had it for a while. Scientists probing the normal species of hunger have recently zeroed in on centers in the brain devoted to its control.

Don Quixote called hunger "*la mejor salsa del mundo*," the world's best sauce. The *Oxford English Dictionary* defines it as "that uneasy or painful sensation caused by the want of food." Hunger often involves an aching or growling stomach, but it can also elicit weakness, dry mouth, and—sorry, Dottore Benedetti—headaches and loss of concentration. Hunger pangs peak at midday, even in the absence of external time cues. (A pang, incidentally, is quite different from a growl or rumble, known as borborygmus. The latter arises from the muscular activity of the stomach and small intestine, whether empty or full—the noise is just louder when there's no meal to muffle it.)

It was once believed that the drive to eat originated solely in the stomach. However, the great nineteenth-century neuroscientist Charles Sherrington observed that hunger persists even in those who have had their stomachs surgically removed. Now the finger is definitely pointing elsewhere in our anatomy. A recent neuroimaging study found plenty of "disturbance" in the brain in response to hunger. Oddly enough, the areas of activity differed somewhat in men and women. Researchers at the National Institutes of Health used PET scans to look at brain activity in twenty-two men and twenty-two women after they had fasted for thirty-six hours and again after they consumed a liquid meal to satisfy their hunger. When they were fasting, all of the subjects showed more abundant blood flow in the hypothalamus, a brain region known to regulate the basic physiological response to hunger. But the hungry men showed more activity in the paralimbic areas of the brain, which govern emotion, than did the famished women; and when sated, in an area of the prefrontal cortex associated with processing reward. This suggested to the researchers that men may experience more reward from eating than women do. I doubt this, though my experience is admittedly limited.

Hunger is one thing; appetite, the desire to eat, quite another. While the two often coincide, we all know that appetite can easily occur in the absence of hunger. Many of us have the desire to eat long before our stomachs feel hollow because food looks or smells good. Or because it's noon and time for lunch and someone has served us up a plate of broiled trout. Or because we're bored and want the stimulus of hazelnut cake. What is the physiological loop that translates into the craving for crisp pita chips and hummus with a light sheen of olive oil?

What we have learned about the biology of appetite in the past ten years pours into a large pot indeed. You may have thought you were in control of your own food impulses at a given moment, but new science suggests that a complex mix of chemical messengers is in fact dictating your dietary decisions.

When endocrinologists at Harvard reviewed all the molecules that regulate appetite, turning it up or down, signaling "eat" or "don't eat," they found dozens of chemical messengers lurking in mouth, stomach, intestines, liver, bloodstream. Some of these couriers act rapidly, from meal to meal, controlling appetite and satiety for any single dining experience. Others exert their effect over the long term, keeping track of the body's fat supplies and telling the brain when they're running low so that it can step up appetite. The long-term signals may spur the production of short-term "I'm hungry" messages or quash them. You're probably unaware of these chemical fluctuations, but they direct your behavior, either driving you toward that lunch buffet or letting you get on with your work.

Two brain regions read this soup of signals, and a sophisticated crosstalk between them determines the outcome. The recipient of the short-term signals for any particular meal is the hindbrain, or caudal brainstem. The umpire of signals relating to the long-haul need for food is the hypothalamus, especially an arc-like cluster of five thousand or so neurons known as the arcuate nucleus. As early as 1912, postmortem exams of very obese subjects revealed lesions in the hypothalamus, suggesting that this part of the brain might be important in regulating appetite. Researchers lately confirmed that the arcuate nucleus integrates and adjudicates the sometimes conflicting messages from a staggering array of hormones, nutrients, and nerves to decide which way to make the body feel, like eating or not. It also determines how to adjust metabolism—the set of chemical reactions by which the body extracts energy from food or stores and uses it for all of its activities—to shift it up or down, to waste or conserve energy.

One of the star players among the "hormones of hunger" is ghrelin (pronounced GREL-lin, from the Old English *ghre*, to grow), a small peptide secreted primarily by the stomach and upper small intestine, which acts on the brain as a potent appetite stimulant. Volunteers injected with ghrelin get very hungry and eat 30 percent more than they normally would.

David Cummings and his colleagues at the University of Washington see ghrelin as a "saginary" hormone (from the Latin *saginare*, to fatten)—the product of a thrifty gene that evolved to help animals consume and store fat well, thereby increasing their chances of survival during times of famine. When the researchers measured circulating ghrelin thirty-eight times during a twenty-four-hour period, they found that levels of the hormone rise and fall dramatically over the course of a day. Before meals, levels go up precipitously, by nearly 80 percent, peaking when the stomach is empty, just before each meal, and then plummeting to trough amounts an hour after eating.

"The empty stomach, however, is not the trigger for pre-meal ghrelin surge," says Cummings; instead, it's the brain anticipating a meal. If you're accustomed to having four meals a day at regular times, you'll experience four ghrelin spikes, one before each expected meal. If you're used to two meals, there will be two spikes. As the number of scheduled meals decreases, so does the number of surges, but the size of each surge increases, as does the sensation of hunger and the amount of food eaten at each meal.

Certain hormones oppose the actions of ghrelin, among them leptin. Not long ago, this hormone made a big splash in the press as a possible magic bullet for treating obesity. Produced by fat cells, leptin is made and released into the blood in proportion to the amount of body fat one is carrying; from the blood it travels to the hypothalamus, which responds by modulating appetite and metabolic rate. The more fat you possess, the more leptin is made by your cells. Leptin appears to be the body's way of telling the brain whether fat stores are sufficient so that it can match caloric intake to caloric expenditure—a feat it's remarkably good at: For most people, intake exceeds expenditure by less than 1 percent. (Even this small difference, however, can lead to added pounds over the long term.)

When leptin levels fall, the brain reads this as a warning sign of deprivation and sends out signals to the body to ratchet up appetite and make metabolism more efficient, reducing the expenditure of energy until the lost weight is recovered. With weight loss, and the accompanying decrease of leptin levels, the hypothalamus sends neural signals to the hindbrain to render it less sensitive to short-acting satiety signals from the gut, explains David Cummings. "Consequently it takes more food at a given meal to make a person feel full, so one tends to eat more at each meal until the original body weight is regained. In this way, the long-acting signals ultimately govern food intake at individual meals." Indeed, this is what makes dieting, and especially keeping weight off, so difficult. The body has this sophisticated mechanism for protecting against weight loss.

Leptin has worked as a therapy for obesity only in rare instances, for people who genetically lack the hormone. In other cases of obesity, resistance to leptin may develop, and raising levels further is not very effective. Still, it is one powerful hormone. Studies in mice suggest that during neonatal development, leptin molds the circuitry of the brain, strengthening pathways that suppress appetite and weakening those that stimulate it. Taking in too much or too little food at this critical stage of infancy may actually change the way the circuits are shaped, affecting appetite and how the body responds to fat in adulthood. In fact, say researchers, leptin's molding of the appetite circuitry early in life may be the biological underpinning of what's known as body-weight set-point—a kind of memory for the weight range the body wants to maintain throughout life. You can move around in this set-point range through diet and exercise, but you can't change its parameters.

So here's an angle on appetite: Your desire for an early lunch on a Wednesday in June may ultimately have its roots in the distant days of infancy.

Whatever its origins, your body's cry for food will not be silenced, so you suggest a lunch break and hop in a taxi with colleagues to go to a nearby salad bar. What do you choose? Fresh greens? Fried chicken? Marinated tomatoes with fresh mozzarella?

What we opt to eat, and why, is nearly as complex a matter as appetite itself. Experience, childhood associations, legacies from our deep past, all play powerfully into food selection. The underlying drive for certain sweet, salty, and umami flavors is rooted in the need for basics, for calories and essential nutrients. The sour we pick only selectively, avoiding the strong acidity or tartness of unripe or spoiled fruit. The bitter we shun, with good reason. In perusing that salad bar, I would pass quickly over the potatoes with fresh mint—a learned aversion. Some twenty years ago, my husband made a potato salad with mint, garlic, and olive oil. Unfortunately, the potatoes he used were spoiled, and the strong dressing disguised the bitter taste of solanine, a poisonous alkaloid made when potatoes "green" in response to overexposure to light. I ate the salad with gusto and ended up sick as a dog. Two decades later, I still can't contemplate revisiting this dish.

Nausea—hunger's antithesis—is a potent protective tool. Just what causes the sick head and roiling stomach remains an enigma. But most of us have experienced the sensation at one time or another, from a bad batch of tuna, too much alcohol, tobacco, or salt water, illness, disgust, unpleasant smells, medication, pregnancy, or motion sickness (the word "nausea," in fact, derives from the Greek *naus*, for ship). So powerful is the sensation that mothers often remember the discomfort of morning sickness long after they've forgotten the pain of childbirth. As nausea goes from bad to worse, the salivary glands flow, the heart races, blood pressure drops, blood vessels in the skin constrict, and we grow pale and clammy. At the same time, electrical activity in the stomach shifts, causing its muscles to relax. The esophagus contracts, the upper small intestine empties its contents into the stomach; then, in one giant contraction coordinated by the brain, the abdominal muscles and diaphragm squeeze, exerting pressure on the stomach lying between them, and the retching begins.

Of course, more pleasant associations may also dictate your choice of lunch foods. Most of us lean toward the familiar. The food that sustained my family was a mix of Jewish, 1950s American, and German cuisines: matzo brei, meat loaf, bratwurst (those stout pork sausages that release down the chin a stream of hot, spicy juice), and, with the arrival of my adopted sister from Seoul, a touch of the exotic: Korean beef and kimchi, a dish that elicits the intense oral burn that some find so pleasurable.

Roasted chicken is my comfort food, soothing in part because of its intimate link with my grandmother. "Eat," she would say, "eat," as she thrust another slab of tender white meat on my plate in her tiny Upper West Side apartment. When I pleaded a full stomach, she would wrap in brown paper the leftover pullet, complete with its pan juices, and stuff it in my briefcase for my plane trip home. This fragrant, dripping package I would tuck into a corner of the overhead baggage compartment, whence wafted aromatic vapors of rosemary and garlic to torture my fellow travelers. I did this culinary shipping trick not out of a sense of obligation or duty but because I relished the sweet flesh, so tender it seemed to evaporate in my mouth. Grandma's magic touch with poultry she passed on to her son, my father, who would dish out to his flu-stricken daughters loving bowls of hot chicken soup—Jewish ampicillin—for me, the flavor of familial love.

There may be more to comfort food than simple familiarity. Some kinds of fare contain substances that appear to boost mood. Foods such as cold-water sardines, tuna, salmon, and walnuts, loaded with omega-3 fatty acids, may have an important impact on how we feel. In 2005, William Carlezon and a team at Harvard discovered that, in rats anyway, these compounds work at least as well as prescription antidepressant drugs at lifting mood. A likely explanation for this effect is the positive impact the compounds have on the brain's mitochondria (the energy-producing power plants of all body cells), which may ultimately enhance communication among neurons in key areas of the brain. However, Carlezon emphasizes that it took a month of feeding rats an omega-3-enriched diet to see these effects. "Shorter treatment periods were not effective," he says. "So an occasional piece of fish won't do it—you need a sustained change in diet." Carlezon's finding bolsters earlier research showing a correlation between fish consumption and lower prevalence of major depression. "This work provides more evidence that our behavior—including the selection of the foods we use to fuel our body—can have a tremendous influence on how we feel and act," he says.

Another study suggests that some fare may not only boost psychological comfort but relieve physical discomfort. Researchers have found that foods rich in butter, oil, and other kinds of fat can reduce the perception of pain. Subjects fed a meal of pancakes loaded with cream and melted butter ninety minutes before having their forearm immersed in freezing water reported feeling less pain than those who had eaten pancakes of equivalent calories, but made with skim milk and water. The greatest pain relief occurred about an hour and a half after the meal. Because a liquid meal failed to offer the same relief, the scientists suspect that the effect may depend on so-called orosensory stimulation—smelling, tasting, and feeling those rich fatty pancakes—which may trigger the body's natural painkilling opioids.

Chocolate, known for its uplifting effects, may work its magic by the same method, sparking such a chemical kick in the brain that feels good. One study suggested that eating chocolate may create a positive mood not just in the one who's indulging, but also—if she's pregnant—in her baby. When researchers at the University of Helsinki looked for a link between the amount of chocolate eaten by pregnant women (especially those feeling stressed) and the behavior of their babies, they found that babies born to women who ate chocolate daily during pregnancy were rated more active, more likely to smile and laugh, and less fearful than the babies of those mothers who didn't indulge.

For flavor or familiarity, comfort or craving, you've selected your meal, perhaps egg salad with greens and a thick wedge of chocolate pie.

Take a bite of that pie. The mouth is full of food sensors, and not just those devoted to taste. As you tuck into creamy chocolate and buttery crust, the highly sensitive receptors in and around your teeth help to modulate the secretion of saliva, a fluid made of 99 percent water and 1 percent magic—magic in the form of sodium ions, enzymes, and a host of other organic substances, among them bacteria-fighting mucins, without which our teeth would decay. Special mechanoreceptors on your tongue sort the pieces of your mouthful by size so as to place large, tough ones between the teeth for chewing. Inside the tooth and in its sockets are still other sensors—thousands of nerve endings, the highest density in the body—which are there not for the purpose of enhancing toothache or the pain of the drill, says Peter Lucas, an anthropologist at George Washington University, but to offer fine-scale detection of forces. This helps with the up-front decisions we make about the taste, texture, and quality of food, and whether or not to swallow.

Take a look at your teeth in the mirror. The shiny white enamel crowning each tooth is the strongest tissue in your body, and necessarily so. According to Lucas, our jaws put as much as 128 pounds of pressure on our teeth when we chew, creating tensile forces to crush, grind, slice, and break apart food into particles. All of this pressure, or mechanical loading, is important not just for grinding food but for maintaining bone in our jaws; without it, the bone would slowly shrink over time. Remove a tooth, and hence reduce the pressure of chewing, and the jawbone in that area will decrease by 25 percent.

Glance again at those pearly ones. Chances are they do not shine as stellar examples of ideal dentition. By animal standards, human teeth are extraordinarily disordered and the only part of the body that requires regular surgery. We may have both evolution and diet to thank for this. Because tool use and cooking have reduced our food to small particles or mush, like egg salad and mashed potatoes and chocolate pie, we don't chew nearly as much as our ancestors did. On average, we spend only an hour a day chewing (a sixth of the time a chimp spends for the same calorie intake). And even during that solo hour, with our diet of soft and processed foods, we don't generate much force. Compared with a raw potato, a cooked one reduces stress to molars by more than 80 percent.

Chewing, or the lack thereof, can quickly transform the anatomy of our jaws, says Dan Lieberman, a biological anthropologist at Harvard. When Lieberman fed a soft diet of cooked food to small furry animals called hyraxes, or rock rabbits, he found that their snouts developed thinner, shorter bones than hyraxes fed a diet of raw food. In Lieberman's view, something similar has been happening to our tribe.

"Since the Paleolithic, our faces have been reduced in size by about 12 percent," says Lieberman, "and most of this shrinkage has occurred in the mouth and jaw." Our teeth, on the other hand, have largely maintained their number and size despite this facial diminishment, causing crowding and other dental ills.

Even with the help of saliva and plenty of chewing, swallowing is no easy feat. I grasped this for the first time while watching what transpired inside the pink gullet of a medical student at the University of Virginia School of Medicine. An otolaryngologist had numbed the young woman's throat and inserted through her nose a fiberoptic tube with a camera attached, which projected the image on a giant movie screen.

"You're looking at Lisa's pharynx," said the anatomist in charge, Dr. Barry Hinton. This is the cavity where the hollows of the mouth and nose join in the back of the throat, familiar to all who have experienced postnasal drip. On the big screen, the pharynx looked for all the world like a pulsating pink cave. Dr. Hinton asked Lisa to breathe normally as he pointed out the details of her larynx, also known as the voice box, the organ that plays so crucial a role in breathing and speaking: its opening, or glottis, and its little fold of vocal cords, which widened and narrowed with beautiful proficiency as she inhaled and exhaled. Here the channels for air and food diverge, leading, respectively, to the trachea, passage to the lungs, and to the esophagus, tunnel to the stomach.

"Speak a little, if you can," Dr. Hinton said. Lisa achieved her mission with some difficulty, gagging at first, then mastering the task, the window in her glottis narrowing and opening wide with the voicing of the *p* in *please* and the *t* in *take*, as in "Please take out this tube."

"One final task," said Hinton. "Swallow." Lisa grimaced. Then the coral circles of muscle in her throat contracted in a quick spasm, raising the larynx and swinging the epiglottis into place over its opening to shut off the respiratory tract so that she might swallow without choking. It seemed nothing less than wondrous.

You're slowing down on your pie now, nibbling here and there as your appetite tails off. The human stomach expands to receive a meal of as much as two and a half pints (that's about half the capacity of a dog's stomach, and just a hundredth that of a cow). This it holds for a few hours, depending on the amount of food, before delivering it, by degrees, through waves of contraction, to the small intestine.

Stretch receptors in the stomach help to signal fullness. But the matter is not so straightforward: At least half a dozen messages from the stomach and intestines reinforce the "stop eating" message. Two hormones, CCK and PYY, made by intestinal cells and secreted in response to the presence of food in the gut, play a key role in delivering the satiety signal to the brain. Give people an infusion of these hormones, and they'll cut back on food intake and end their meal earlier. In one recent study, people injected with PYY and then offered a free-choice, all-you-can-eat buffet two hours later ate a third fewer calories than people who were injected with a saline solution; these appetite-suppressing effects lasted twelve hours.

How quickly you feel satisfied also depends on what you eat. Foods are not equally effective at suppressing hunger signals. Those rich in fiber, which move more slowly down the gut, may trigger more PYY than do fast foods made of refined carbohydrates, which are quickly dissolved in the stomach. David Cummings and his team have shown that protein and sugar both suppress ghrelin, triggering a quick 70 percent decrease in the hunger hormone, while fat makes ghrelin levels fall more slowly and only by about 50 percent. The researchers suggest that this weak suppression of ghrelin by high-fat foods could be one of the mechanisms underlying the weight gain that comes with high-fat diets.

Whatever the content of the cuisine, however, the message eventually gets through: *enough.*

5

POST-LUNCH

●

THE SUN IS HIGH, the breeze light, the meal heavy in your stomach. Best to walk the mile or so back to the office. As you stride along the sidewalk, threading through the crowd, you're setting in motion more than fifty bones in the ankle and foot—a quarter of the bones in your body—as well as multiple muscles and ligaments, all interacting dynamically with the ground.

"If I could not walk far and fast, I think I should just explode and perish," wrote Charles Dickens. Goethe composed poems while walking. So did Robert Frost and Dante. Some observers have even credited walking, legs and arms swinging in pendulum time, with imparting rhythms for famous poems and prose, including Dante's *Purgatorio,* with its measures that mimic the human gait.

Whether or not we need walking for sane mind or sound meter, we do seem built for it. To fathom what's going on in the body during such a seemingly simple act of human locomotion, scientists analyze the movement of limbs and expenditure of energy of subjects walking or running on a treadmill. As a volunteer subject in one such experiment in 2005, I was wired up and put through the paces at Dan Lieberman's laboratory at Harvard University. On my feet were pressure sensors to show my heel and toe strikes. Electromyography sensors revealed the firing of my muscles, and accelerometers and rate gyros on my head detected its pitch, roll, and yaw. Small silver foam balls attached to my joints—ankle, knee, hip, elbow, shoulder—acted as infrared reflectors for three video cameras mapping in three-dimensional space the location of my limb segments. Later I would wear a mask connected to equipment that gathered information on how much oxygen I consumed while walking and running, a measure of my energy expenditure.

All this gear was about as comfy as a hair shirt, especially the head part, improvised from elastic, foam, and wires. But learning the lore of locomotion was worth the discomfort.

Walking feels easy because it easily converts the body's potential energy to kinetic energy, Lieberman explained. A walking human body is not unlike an inverted pendulum. The body pivots over a relatively rigid or stiff leg, with little need for energy input—the potential energy gained in the rise roughly equals the kinetic energy expended in the descent. By this trick the body stores and recovers so much of the energy used with each stride that it reduces its own workload by as much as 65 to 70 percent.

Watching on a computer screen the tabulated results from the experiment, I had to marvel at the ingenuity of the moving body, the clockwork firing of the muscles, the regular pumping action of arms and shoulders, the consistency of our long-legged stride. Walking is a highly efficient form of locomotion for our species—at least at optimal speed. Around 4.2 feet per second, or a little more than 3 miles an hour, is most economical, says R. McNeill Alexander, a biologist at the University of Leeds, in part because muscles work best at the stride length and frequency characteristic of this pace. As pace increases or decreases from this optimum, the cost to the body rises rapidly. But somehow the body knows how to minimize its costs, even when it's forced to move in awkward strides. In one study, Canadian scientists asked athletes to tread with weird gaits—little mincing steps or strange plodding strides. They found that the athletes automatically compensated for the odd gaits and minimized their energy expenditure by adjusting their pace and stride frequency. When we walk, say the researchers, the relationship between our speed and our stride length and frequency is not an accident of mechanics. The body is monitoring gait all the time and making necessary adjustments, all without a conscious thought from us.

But now you're running a little late, so you pick up your pace. As you cast aside energy efficiency in favor of urgent speed, your breathing grows more labored. At rest, you breathe in and out about sixteen times a minute, inhaling some eight quarts of air. But switch into high gear to hustle back to the office or sprint across a busy intersection and your need for air will jump fifteen- or twentyfold. How does the body know when it's low on oxygen and needs to breathe harder?

For more than a century, scientists have been searching for the elusive oxygen sensor. Not long ago, biochemists at the University of Virginia discovered a likely candidate in a type of nitric oxide known as SNO. Nitric oxide is the gas generated during a lightning storm and best known for reacting with ozone to cause smog. The body, it turns out, makes nitric oxide in its own cells for a range of purposes, from controlling the muscles of the gastrointestinal tract to dilating blood vessels. Now it's believed that the SNO form of nitric oxide is also the messenger that allows the blood to communicate with brain regions that control respiration.

I love this idea that a gas born of lightning also sparks the heavy breathing necessary to carry us back to the office on fleet feet.

A little breathless but invigorated by your walk, you duck into the bathroom to freshen your breath with a quick tooth brushing. Here's a little-known fact to enliven your dental scrub: Brushing is not a simple matter of scouring the scum from your teeth; rather, it's an experiment in social evolution, suggests Kevin Foster, a biologist at Harvard. Like it or not, your mouth is home to a teeming society of bacteria that occupy distinct niches on tongue, teeth, and gums. "Brushing may mix bacteria that were previously surrounded by their clone mates with unrelated bacteria from another part of your mouth," says Foster. This mixing affects the evolution of their communities, which in turn determines whether they cause troubles such as tooth decay and bad breath.

That the maw is neighborhood to a secret microscopic life was first discovered by the seventeenth-century Dutch draper and naturalist Anton van Leeuwenhoek. One day, in a characteristic moment of curiosity, Leeuwenhoek scraped a little plaque from his teeth and put it beneath his microscope. He saw "with great wonder ... very many little living animalcules, very prettily amoving [which] hovered so together, that you might imagine them to be a big swarm of gnats or flies."

Only lately have we learned that the mouth hosts truly cosmic microbial communities, easily exceeding in number the six billion or so people on Earth. (Consider this: In one slow kiss, partners swap more than five million bacteria.) The six hundred or so different species of oral occupants are not uniformly distributed or haplessly floating about, but flourishing in organized communities that adhere together in "biofilms" and settle into specialized niches. These biofilms protect the bacteria and encourage their growth in family groups. The so-called Red Complex, for example, is an alliance of three species that appears to contribute to gum disease. Brushing disrupts these social relationships, says Foster, inhibiting their ability to grow, thrive, and rot your teeth, irritate your gums, or promote halitosis.

Research suggests that bad breath is mainly the result of these tiny mouth microbes satisfying their taste for proteins. In digesting the proteins, they produce what one microbiologist, Mel Rosenberg of Tel Aviv University, calls a bouquet of "truly fetid substances": hydrogen sulfide (that rotten-egg odor), methyl mercaptan and skatole (which produce the odor in feces), cadaverine (the smell of rotting corpses), putrescine (the odor of decaying meat), and isovaleric acid (the stench of sweaty feet).

Perhaps the world's expert on breath odor research and a self-described smell-ologist, Rosenberg developed a clinical test for bad breath called a Halimeter and a user-friendly litmus test called the OK-2-Kiss test, which measures the presence of troublesome bacteria and malodor. Rosenberg lists twenty-two species of bacteria known to cause bad breath. Normally, saliva washes away both bacteria and their stinky metabolic products, he says, but sometimes saliva doesn't reach the back of the tongue, where bacteria can hide and "putrefy" postnasal drip. A mouth dry from a long night of mouth breathing or a morning of fasting can worsen the situation. So can too much talking. (It's a particular scourge for politicians.) However, Rosenberg does not advise trying to do away with oral bacteria. Some species play an important protective role, he says: When their populations are reduced, say by the chronic use of antibiotics, the tongue becomes prey to colonization by candida, a yeast-like organism that causes disease.

So how to avoid the dreaded halitosis? According to Rosenberg, Italians chew parsley. Iraqis gnaw on cloves; Brazilians, cinnamon; and Indians, fennel seeds. Thais munch on guava peels, and the Chinese drink rice wine with crushed eggshells or eat persimmon or grapefruit or red dates. If you're without access to such spices, herbs, or fruit, Rosenberg recommends keeping your mouth moist and brushing and flossing after meals, especially after eating foods rich in protein.

• • •

Back at your desk with a full stomach and relatively fresh breath, you're ready to tackle that stack of papers, organize the afternoon, supervise staff. You've forgotten all about your egg salad. Fortunately, your body hasn't. It's just starting the business of digestion, overseeing the millions—no, billions—of obscure little laborers managing the hard toil of eggs, salad, and pie, quietly, invisibly, so you are able to think of other things.

The clandestine events of digestion were long ago described by William Beaumont, who gained an excellent view of the subject thanks to the strange misfortune of a nineteen-year-old Canadian trapper named Alexis St. Martin. Beaumont, a U.S. Army surgeon, was called one June morning in 1822 to treat St. Martin for a large wound in the abdomen. The poor trapper had found himself at the wrong end of a shotgun. The gun had fired accidentally and struck him at a distance of only three feet, "literally blowing off integuments and muscles of the size of a man's hand," wrote Beaumont. The gaping wound was such that the trapper's death seemed certain. Despite great loss of blood and days of high fever, St. Martin survived. But the injury left a permanent hole in his stomach, a kind of valve the size of a forefinger that had to be plugged so food wouldn't ooze out during meals. The hole allowed Beaumont to see inside St. Martin's stomach to a depth of five or six inches and to conduct more than a hundred groundbreaking experiments on the workings of the stomach, its secretions, and the process of digestion.

"Pure gastric juice . . . is a clear, transparent fluid; inodorous; a little saltish; and very perceptibly acid," wrote Beaumont. "It is the most general solvent in nature . . . even the hardest bone cannot withstand its action." It's true. The gastric juice sloshing about in your core is one powerful brew, made of pepsin, an enzyme that breaks down the proteins in food, and hydrochloric acid—a substance so caustic it can demolish bacteria and dissolve iron—which provides the acidic environment that pepsin requires to do its work. Smelling or tasting food, or just thinking about it, stimulates cells in the lining of the stomach to secrete hydrochloric acid. Among the stomach's more famous feats is its ability to digest, say, boiled beef, with the help of this acid without burning up its own tissue or digesting itself—a talent it owes to its inner walls, which possess a layer of mucus and bicarbonate that shields it from its own corrosive contents. When gastric juice leaves the protected environment of the stomach and backs up into the esophagus, the result is the painful sensation of heartburn. If occasional, this backup is only bothersome, but if frequent, it's dangerous, as gastric juices can erode or destroy the lining of the esophagus. The production of these juices is lowest in the morning and peaks from about 10 P.M. to 2 A.M., which explains why peptic ulcers act up and heartburn flares during these hours.

Despite its special equipment, your stomach is dispensable. An effective storage facility and preparer of food for digestion, kneading it into small particles, pulverizing and sterilizing it, the stomach otherwise plays just a small part in the actual process of digestion and virtually no part in absorption (except of certain drugs, such as alcohol and aspirin). The work of absorbing takes place through fingerlike projections in the intestines called villi.

These days, detailed study of digestion no longer requires a bullet hole; with special scopes and chemical tools we can observe in the murky fundus or dim niches of the duodenum, even in the tiny villi, events occurring at the level of individual cells and molecules. We can track these activities over time, listen to signals flying to and from the gut, and gawp at its unexpected "intelligence."

That we digest our meals without taxing the brain is largely due to an independent, self-sufficient "brain within the belly," according to Michael Gershon of Columbia University. The brain in the head controls what goes on at the top and bottom of the digestive system, but what happens in between is managed primarily by what Gershon calls the "brain gone south."

Inside the thirty-two-foot tube of your intestinal tract lies an intricate web of millions of nerve cells that runs things, controlling both the movement and the chemistry of digestion. Only in the past few years have scientists begun to unravel the secrets of this intelligent network, known as the enteric nervous system. Gershon was among the first to suggest that the system was driven by the very same chemicals that transmit instructions in the brain. He and others have found at least thirty brain chemicals of different types operating as messengers in the bowel. These chemical messages allow the enteric nervous system to perform a plethora of tasks without the help of the brain—from sensing nutrients and measuring acids to triggering the waves of motion that propel food along the digestive tract and coordinating with the immune system to defend the gut.

According to Gershon, a steady stream of signals flies back and forth between the two "brains." Think of those butterflies in your stomach before you delivered your presentation. "We all experience situations in which our brains cause our bowels to go into overdrive," says Gershon. But as it turns out, the message traffic is heavier going north, from midriff to mind, by an order of about nine to one. "Satiety, nausea, the urge to vomit, abdominal pain, all are the gut's way of warning the brain of danger from ingested food or infectious pathogens," Gershon explains.

Your intestinal tract is surprisingly smart, versatile, brain-like. But its ultimate achievements are not yours alone: Your resident bacteria play a far larger role in digestion than ever imagined.

You may have been a sterile, singular being in the womb, but once you entered the birth canal and then the world of nipples and hands and bed sheets, you picked up an ark of microbial handmaidens. Soon the little buggers were everywhere, like words filling a page, in folds of skin, in orifices of nose and ears, and especially in the warm, cozy tunnels of your digestive tract, from mouth to anus. By the age of two, "the human body is grossly contaminated with microbes," notes David Relman, a microbiologist at Stanford University. In fact, "of all the cells that make up the healthy human body," he says, "more than 99 percent are actually microorganisms living on the skin, in the gut, and elsewhere." The small intestine is densely settled, with 100 million bacterial cells per milliliter (.06 cubic inch); the large intestine, or colon, 100 *billion* per milliliter. The total weight of all these bugs has been estimated to be more than two pounds.

In 2005, scientists for the first time attempted to enumerate the different microbial species inhabiting the gut. Microbiologists used genomic sequencing to take a census of the gut flora of three healthy adults and discovered close to four hundred species, more than half of which were entirely new to science. The researchers suspect that this is the tip of the iceberg, that the number of species of gut microbes may be closer to six or seven thousand. Hundreds of these species bring with them genes that endow us with traits and functions that are useful to us—and that we have not had to develop on our own. In this way, they expand our own genome and act as master physiological chemists in our bodies. In fact, say the scientists, your body can best be thought of as a kind of genetic superorganism, a rich amalgam of human and microbial genes.

My bugs are probably different from yours. Studies of twins and their marital partners suggest that our genetic makeup helps determine the types of bacteria that are attracted to our alimentary tracts and set up shop there. But a host of environmental factors also play into the picture: where we live, what we eat and drink, our hormones and our hygiene. The bacteria we meet as infants help shape the populations we'll carry throughout life. Babies born by cesarean section may have different species than those born through the vagina. (Baby mice, at least, swallow the bacterial bits and pieces floating around in the birth canal as they make their way down.) Breast-fed babies tend to be colonized by bifidobacteria, and generally have fewer gut troubles than formula-fed infants, who are inhabited by more clostridia, bacteroides, and streptococci. Early use of antibiotics, too, can deeply affect the populations.

As long as our microbial communities remain stable, we coexist in peace. The load of bugs is potentially dangerous, but the dense competition among them usually keeps any one player from dominating. Also, the possibly destructive ones are held at bay by the body's immune cells, which come to know the resident bacteria, learn to neutralize the toxins they make, and mount an effective attack if the interlopers venture beyond the walls of the digestive tube. However, if something changes the composition of this intestinal community —say, a bacteria-laden piece of fresh fruit or vegetable, eaten in a place where the local microbes differ from those at home—there may be unpleasant consequences.

Once while traveling in Guatemala, I succumbed to a craving for salad at a hotel restaurant and nibbled a bite or two of fresh lettuce and tomato. Not long thereafter, I lay in a feverish sweat in my hotel room, every few minutes stumbling to the bathroom, a victim of turista. (After twenty-four hours of agony, just as a candlelight Christmas procession passed by the window, I sat up in bed fully recovered—which my Catholic husband took for a miracle and I chalked up to a well-tuned immune system.)

Most of us have suffered in similar fashion. We endure a misbehaving bowel until—miracle or not—the immune system learns the nature of the new bacteria.

A much more serious disruption can result from the use and misuse of antibiotics. Such environmental meddling may create an imbalance in the normal bacterial consortia, wiping out some of the gut's residents and allowing a single strain—often a pathogen such as *Clostridium difficile*—to multiply. Even worse, in the dense, gene-swapping microbial communities of the gut, it may also encourage the evolution of microbial pathogens resistant to antibiotics.

But many of our abundant microscopic denizens are neither potential troublemakers nor passive bystanders, says Jeffrey Gordon of Washington University: "They're companions essential to our digestive well-being, symbionts that have coevolved with us and benefit from the association, just as we gain from our alliance with them." For years we've known that friendly microbes, or commensals, help us make vitamins and establish tight-knit communities that keep out potential pathogens. They also metabolize nutrients so that we can absorb them more readily (especially such otherwise indigestible components as plant cell walls). But how they accomplish their good works has been poorly understood. Most of these microorganisms are fiendishly difficult to study. It's hard to keep them alive outside the gut. And even if scientists could sustain them in the lonely isolation of a petri dish, bacteria in culture probably wouldn't behave the same way they do in their normal ecosystem in the intestines.

Gordon realized that the only way to get real insight into these beneficial bacteria is to study them in their natural setting. So he and his colleagues have developed an ingenious approach. In germ-free plastic bubbles they raise germ-free mice, which have none of the trillions of microbes that would normally reside in them. Then they introduce common gut bugs one at a time and study their effects.

What they are learning is revolutionizing our view of ourselves and how we process the food we eat. Without our resident bacteria, Gordon has discovered, our intestines would not grow properly. One way the gut protects itself from natural toxins and its own powerful acid secretions is by shedding its own lining every week or two. As the replacement cells mature, they travel from the base to the tips of those little finger-like villi lining the intestines. They do so, Gordon has found, only with the help of bacterial signals, which ensure their healthy development. Without these microbial messages, our intestines and their all-important villi would fail to grow normally.

Gut bacteria also protect this intestinal lining. Scientists at Yale discovered that the bugs help to activate the body's machinery that repairs injured cells. In killing off our friendly bacteria, antibiotics can inhibit the processes necessary for this protection and healing. Moreover, certain bugs help us tolerate harmless food proteins and other innocuous foreign matter floating inside the alimentary tract. If our immune cells react to these, triggering inflammation, it's bad news for us. One prominent microbe with the cumbersome name *Bacteroides thetaiotaomicron* ensures that our immune systems leave alone these innocent interlopers.

But here's the real shocker: Our *B. theta* and other bacteria may also help determine our girth by influencing how many of our calories are transformed into fat. Gordon and his team have found that germ-free mice can eat 29 percent more food than mice with normal microflora and still maintain a svelte figure, with 42 percent less body fat. Adding a community of gut bacteria to the intestines of the germ-free mice caused them to increase their body fat content by 60 percent in two weeks, even though they didn't eat any additional food. "That's because these bacteria improve the efficiency of calorie harvest from the diet and help the body deposit the extracted calories in fat cells," Gordon explains. When he and his colleagues probed the genome for *B. theta,* they found that many of the microbe's genes are dedicated to processing carbohydrates that we don't have the genes to digest. Without bacteria such as *B. theta,* the carbs would simply pass through our system without caloric gain.

Recently, Gordon and his lab mates took their experiments a step further. In comparing the gut bacteria of fat and lean mice, they found that fat mice had a larger proportion of a type of bacteria called Firmicutes and a smaller proportion of Bacteroidetes. When they transplanted the Firmicutes-rich microbial mix from fat mice into germ-free mice, the recipients put on more body fat than those receiving a mix of microbes from lean mice. In human studies, the team found that similar Firmicutes/Bacteroidetes proportions held true for obese and lean people. And as the obese people in their study lost weight over the course of a year under the scientists' supervision, their gut populations became more like those in the lean people.

"The message from these experiments," says Gordon, "is that the amount of calories available in the foods we consume may not be a fixed value but rather influenced by the nature of our gut microbes." The compositional differences in our resident microbes may affect the caloric density of the foods we eat, and ultimately our predisposition to obesity. Take-home lesson: Consume those nutritional labels with a grain of salt. Depending on your gut bacteria, that doughnut might have more calories for you—possibly as much as 30 percent more—than for your neighbor.

I've come to respect and admire the swarming pool of diverse creatures inhabiting my body. I like to think of them skiddling about in my gut after lunch, freely offering their genetic inventions, sidling up to my villi to whisper words of encouragement to young cells, harvesting nutrients and calories, or just idly spinning circles in the swamp water of my lumen, keeping turista at bay.

How long it takes for bowel, bugs, and brain to digest a meal depends on what we eat and when we eat it. Fat-rich meals take longer to digest than meals rich in protein or carbohydrates. About 50 percent more time is required to empty dinner from the stomach than breakfast —in part because the nighttime velocity of the so-called housekeeping waves responsible for gastric emptying is half what it is during the day.

Other aspects of the gastrointestinal tract also show daily rhythms: the activity of enzymes in the small bowel, the secretion of gastric acid, and the rates at which substances are absorbed in the intestines. Franz Halberg of the University of Minnesota determined that the body processes calories in different ways at different times of day. Eat a single daily meal of two thousand calories for breakfast each day, and you may well lose weight. Eat the same meal at dinnertime, and you'll likely gain pounds—perhaps because the body burns off carbohydrates more rapidly in the morning than in the evening.

If daily rhythms influence how we handle food, the reverse is true, too: Our meal schedule affects the pattern of our circadian rhythms. Scientists have discovered that some of those peripheral clocks in our body depend on feeding time to set their schedules. A pattern of regular meals, three times a day, is the dominant *zeitgeber* for the clocks residing in the cells of our liver, kidneys, and pancreas. This makes good sense from a physiological point of view. The body's major organs have to anticipate the handling of food and water, preparing for the required tasks ahead of time, so they're ready to absorb food, secrete digestive enzymes, and control urine production.

Toy with this regular eating pattern—as some shift workers and jet setters necessarily do—and you may screw up those peripheral clocks, wreaking havoc with your intestinal tract. (One recent study showed that daytime feeding of normally nocturnal rodents completely inverts the schedule of clocks in their peripheral tissues.) This may help to explain why shift workers and jet-lagged travelers, who eat at off hours, frequently suffer digestive upset until they adapt to the dictates of their new schedule.

So, under normal conditions, how long does it take for your egg salad and pie to pass through your inner highways and byways? Studies on so-called whole-gut transit time are few and far between, say scientists, because it's not practical to measure this in large groups in the field. But not long ago, gastroenterologists hurdled the obstacles. In one sampling of 677 men and 884 women in East Bristol, England, participants were persuaded to record the details of their diet and defecations, including careful notations on stool form (using the Bristol scale, from 1, "small hard lumps, like nuts," to 6, "fluffy pieces with ragged edges"). From these records, as well as systematic questioning about bowel habits, the researchers estimated the transit time of meals from food to feces to be fifty-five hours for men and seventy-two hours for women. This may seem like an awfully long time, and it's tempting to doubt the universality of these figures given the usual British diet. But other studies confirm that the average rate is between two and two and a half days.

There is, however, a great deal of variability from person to person and from meal to meal. "A meal is typically a mixture of chemically and physically diverse materials," explains the physiologist Richard Bowen. "Some substances show accelerated transit while others are retarded in the flow downstream." Alcohol consumption tends to quicken transit in both sexes, as does intake of dietary fiber; in women, oral contraceptives slow it. Transit time is generally shorter in older women than younger ones; the change occurs around the age of fifty, which suggests that female sex hormones may have some effect.

Want to speed things up? The safest and most natural way, say the experts, is to eat more dietary fiber.

Food spends only a few hours in the stomach and a few more in the small intestine. After our intestinal cells have done their work, what's left passes on in liquid form to the colon. The remaining dozens of hours are spent here, where water is absorbed—on the order of more than two gallons a day—and wastes prepared for elimination. Because of their bacterial content, the latter must be handled by the body "with circumspection," observes Michael Gershon, confined but also propelled through the colon's sole portal, usually once a day.

• • •

We often think about food. We seldom think about what it becomes. Feces, from the Latin for dregs, are made mostly of water, mucus, bile pigments (which lend stools their brown color), some fats, dead cells, gases, plenty of roughage (primarily undigested cellulose, or fiber, from plant-based foods), quite a lot of bacteria that have lost their grip on the colon, and some 1,200 different types of viruses. Roughage provides most of the bulk. Some kinds of fiber go right through our alimentary tract pretty much intact, offering no calories but a sensation of fullness, as well as exercise for our colon, giving it something to squeeze.

A diet with little roughage will produce about four ounces of excrement a day; one rich in fruits, vegetables, and grains, about thirteen. A diet of meat makes for the strongest smell; of milk, the mildest. The stink in feces arises from skatole (also present in bad breath), a byproduct of the breakdown of the amino acid tryptophan. The human nose is highly sensitive to skatole but does not always find it revolting. In fact, the compound is said to be used in small quantities as a flavoring in vanilla ice cream.

Smell rarely escapes from feces stored in the colon except in the event of flatulence. Breaking wind—farting is the more common term, used at least since the time of Chaucer, who wrote, "This Nicholas anon let flee a fart"—is the release of a bubble of intestinal gases, carbon dioxide, hydrogen, nitrogen, and methane, produced in part by the swallowing of nitrogen and in part by the action of intestinal microbes on food. Most of us fart about once an hour, depending on what we've eaten and whether we're under stress.

Preventing the occurrence is exceedingly difficult. Scientists investigated the phenomenon in a case report on a thirty-two-year-old male computer programmer who experienced extreme flatulence. "Conscious efforts to stifle air swallowing seldom are effective," say the researchers, "and the only 'treatment' alleged to inhibit air swallowing is the prevention of jaw closure by holding an object between the teeth. Our patient . . . put this maneuver to the test. Unfortunately, this treatment was ineffective, as evidenced by 66 gas passages over a 13.5-h period during which he clenched an object between his teeth."

So what about modifying our bacterial flora through the use of antibiotics or by eliminating the fibrous substrate on which they thrive? "We have found that ingestion of a diet in which all carbohydrate is supplied in the form of white rice reduces flatus output," say the scientists. (A rather draconian solution if you consider the limited nutritional value of white rice.) Antibiotics fail to appreciably reduce the problem. Consuming so-called probiotics, live bacteria cultures, to induce a flora that efficiently consumes hydrogen might be useful in theory, they say; however, such "floral modification" has not yet been achieved.

So much for what goes to waste. What's happening to the energetic fruits of your meal encapsulated in the egg salad and pie? A raft of new discoveries has exposed some of the secrets of how your body uses the calories it consumes, shedding light on such mysteries as why your slender colleague Esme can eat anything she pleases and never gain an ounce while plump Phoebe consumes many fewer calories, is constantly dieting, and yet stubbornly retains her excess pounds. If you're more like Phoebe than Esme, there may be a thing or two you can do to shift the balance.

As you pursue the tasks of the afternoon, are you sitting quietly at your desk, pretty much a static lump? Or are you nervously tapping your foot? Pacing the halls? Popping up every few minutes to stretch or find that missing page from your manuscript or get another drink of water? Your fidget factor could be an indicator of your tendency to gain weight.

Just by living, by keeping your heart beating, blood circulating, kidneys, lungs, body cells working, you burn up 50 to 70 percent of the calories you consume, says Eric Ravussin of the Pennington Biomedical Research Center in Baton Rouge, Louisiana. This, your so-called resting metabolic rate, or RMR, is the pace at which the body burns calories while at rest to produce energy to keep its basic functions going. Some 20 percent of daily energy expenditure goes to the brain, about 10 percent to the heart and kidneys, another 20 percent to the liver, and up to 10 percent to digestion.

I recently had my RMR tested at a local health clinic with a portable calorimeter, a relatively new instrument designed to help people with weight problems figure out how many calories they burn daily. "Trying to lose weight without knowing your RMR," the literature told me, "is like balancing your checkbook without knowing how much money you're spending."

The physical therapist at the clinic asked me to breathe into a mouthpiece while the machine measured how much oxygen I inhaled and exhaled. People with a high metabolic rate use up more oxygen because they oxidize (or burn) more calories every hour. I had hoped for a high score, because I thought that people with a high metabolic rate are generally protected from gaining weight.

My rate, however, was a disappointing 1,180 calories a day, quite a bit lower than average. It turns out that metabolic rate is partly determined by body size and composition. Big people usually have a higher rate than smaller people; the more mass you have to move around, the higher your RMR. The physical therapist told me she had seen the full range, from 700 a day in a petite woman in her late seventies to 3,500 a day in a six-foot-five man who weighed more than 400 pounds. On average, a 175-pound man in his thirties burns about 25 calories for every kilogram (2.2 pounds) of weight, which comes out to about 2,000 calories a day. For women, it's typically around 1,400 a day, unless they're pregnant or nursing, which requires an extra 300 to 800. One important factor is the amount of lean muscle mass in your body. Weightlifters burn as much as 15 percent more calories all day long, even in their sleep.

However, the formulas are not so simple. "Though RMR seems to be fixed for a given person," says Ravussin, "there can be large differences even between people of the same sex, weight, and body composition." Why? Scientists are just beginning to unravel the mystery.

A small portion of your calories are burned off daily through thermogenesis, the generation of body heat induced either by exposure to cold or by excess food intake. These days, cold exposure is not much of a factor. "Because humans have evolved behavioral strategies (clothing) to maintain body temperature in cold environments," Ravussin says, "cold-induced thermogenesis accounts for only a small portion of daily energy expenditure."

So-called diet-induced thermogenesis, or DIT, is the body's way of converting surplus calories directly into heat—in essence, wasting energy—and it varies a great deal from person to person. Scientists at Harvard have shown that DIT is under the control of the sympathetic nervous system, which increases activity in the heart, pancreas, liver, kidneys, and other tissues and organs in response to overindulgence. Usually our cells burn only as much energy as they need to. But when we eat too much, the brain may sense the surfeit and activate DIT to burn off some of the excess calories as heat. One of the genes responsible for this neat feat makes a protein that acts as a kind of switch to rev up the amount of energy a cell burns in response to overeating. Variations in this gene may be part of the reason why certain people who pig out never put on an ounce, while others who eat the same amount get plump from the glut.

There may be another thermogenetic explanation for the light/ heavy divide. In one two-month study, scientists at the Mayo Clinic in Minnesota kept at a constant level the food intake and physical activity of a group of subjects, then overfed them by one thousand calories a day. Using state-of-the-art equipment to determine where those extra calories were going, the team found that, on average, a third accrued as fat, a third went to RMR, and a third were burned off by so-called nonexercise activity thermogenesis, or NEAT. This includes all the fidgeting, shifting position, standing, walking, drumming of fingers or toes—in short, all the unplanned physical activity one does in the course of a day.

Overeating didn't stimulate the same amount of NEAT in everyone. Some people fidgeted more in response to the gorging and managed to maintain nearly stable weight; others, who fidgeted less, gained up to nine pounds. This natural fidgetiness of an individual, say the researchers, is probably controlled by genetically determined levels of brain chemicals and can account for a big wedge of calorie consumption, from 15 to 50 percent. That can mean the difference between gaining an extra pound from that surplus pie or burning off these excess calories as you go through the motions of your day.

In 2005, the Mayo Clinic team set out to pinpoint the individual differences in energy expenditure. With precise sensors, the researchers measured the posture and body positions of twenty self-proclaimed couch potatoes during a ten-day period. Half of the subjects were lean, half mildly obese. All wore underwear with embedded sensors to monitor movement every half-second. With the help of this covert window into the subjects' energetics, the scientists discovered that the lean people moved around for two and a half hours more each day than did the overweight people. The difference in activity levels amounted to an expenditure or savings of as much as 350 calories a day.

"When people decide to increase energy expenditure for weight control, they usually include only structured exercise in their calculations," says Eric Ravussin. But the observed difference in NEAT between obese and lean individuals suggests that obesity might be prevented by spending less time in the chair and taking more trips to the water cooler. The researchers do not recommend that we quit our health clubs and exercise programs, just that we note the comparable health benefits of NEAT activities and perhaps step them up. In other words, to get your middle little, don't sit the day away, but rather stand when you have the chance, and by all means squirm, fiddle, twitch, and jiggle.

Afternoon

The afternoon knows what the
morning never suspected.

SWEDISH PROVERB

6

THE DOLDRUMS

I T'S MIDAFTERNOON, when the day, the light, the heat are at their peak—but you, suddenly, are not. For an hour or so after lunch, you were humming along, working on your report, writing that difficult letter, with a fresh mind and full acuity. Now, seeping through your back and shoulders, stealing up your neck to dim your brain, here it is: a slow tide of sleepiness. Your eyelids droop and your blinking intensifies; your face grows slack except for the jaw-stretch of one uncontrollable yawn after another. You give up on the demanding task at hand and slog through the minutes, filling them with whatever mundane tasks may have accumulated during the morning.

It's the doldrums, "where nothing ever happens and nothing ever changes," says Norton Juster in *The Phantom Tollbooth,* "where it is unlawful, illegal, and unethical to think, think of thinking, surmise, presume, reason, meditate, or speculate."

For most of us, the doldrums lie somewhere between 2 and 4 P.M., a dip in the day when the fog of fatigue drifts in to cloud thinking and numb limbs, when we grow inattentive and forgetful and may perform as poorly in matters of manual dexterity, mental arithmetic, reaction time, and cognitive reasoning as if we had quaffed several bottles of beer.

If we lived in Brazil or Panama, we might go home for a siesta (a word derived from the Latin for sixth hour, or the middle of the day). But we have no such civilized tradition, so we struggle through our stupor.

Is this dip—often called the post-lunch or postprandial decline (from the Latin *prandium,* or late breakfast)—inevitable? Or is there some way to avoid the sleepy slide?

This and other questions of weariness, fatigue, rest, and rhythms were the focus of a group of scientists who met not long ago for an annual meeting of the Society for Sleep and Biological Rhythms on Amelia Island in northern Florida. Off the coast, a big storm was brewing; white waves roiled in from a stormy sea, and a warm, gale-force wind thrashed the palm trees and whipped the beach into stinging eddies of flying sand. People gathered their beach umbrellas and blankets and made a run for shelter as a dark mass of ominous clouds moved in from the east.

Inside the well-insulated auditorium of the conference center, however, things were calm and comfortable: the seats deeply cushioned, the air conditioner humming softly, the lights low in anticipation of lecture slides. Soon to speak was Mary Carskadon of Brown University, known for, among other things, devising a system of measuring alertness by testing sleep latency—that is, how long it takes to fall asleep—now the gold standard of assessing daytime sleepiness. Her talk that day promised news on alertness and the sleep-wake cycle at different stages of life.

Despite my eager anticipation of her words, conditions were conspiring against clearheaded attention. On the 7-point Stanford Sleepiness Scale, I must have logged in between a 5 ("foggy; losing interest in remaining awake") and a 6 ("sleepy, woozy, fighting sleep; prefer to lie down"). My mind wandered feebly in the lexicon of lethargy: languor, lassitude, loginess, sluggishness, apathy, stupor, torpor, weariness, drowsiness, sleepiness, and a word I just learned: pandiculation, the act of stretching and yawning.

I was not alone. The man next to me had his eyes closed, head bobbing gently in time with his breathing. Periodically, the sudden blow of chin on chest would awaken him, and he would straighten up momentarily, but then his head dropped again. The woman to my left stifled a yawn; I tried to suppress mine, too—twice—but finally gave in to that satisfying deep inhalation, average duration about six seconds, although it tends to be slightly longer for men.

According to neuroscientists, a yawn can occur alone or in association with stretching and/or penile erection (which may help to explain its duration in men). Its function is still largely a mystery. People once thought it served a role in respiration: Triggered by low oxygen or high carbon dioxide levels in the blood, a yawn was the body's way of trying to take in more oxygen or rid itself of extra carbon dioxide. But when Robert Provine, a psychologist at the University of Maryland, tested this theory by comparing the effect of breathing various gas mixtures on yawning, he found that air rich in oxygen or high in carbon dioxide had no significant effect. Even people breathing pure oxygen feel the urge. Now yawning is thought to be akin to stretching, a way of increasing blood pressure and heart rate and flexing muscles and joints during transitional periods between wakefulness and sleep.

It's also considered a social signal. "Yawns may be a primitive form of nonverbal communication to indicate one's thoughts or mental condition," says Steven Platek of the University of Liverpool. This may suggest why they're contagious. As Dr. Seuss said, it often takes just one yawn to start other yawns off.

Humans begin to yawn in utero, about eleven weeks after conception, but the act becomes contagious only in the first year of life — and only in about half the population. To probe the nature of contagious yawning, Platek and his colleagues conducted a series of experiments to see what might make certain people susceptible.

The team tested sixty-five college students for personality traits that revealed their level of self-awareness and empathy, and then showed them short videos of people yawning, while observing them through a one-way mirror. Just over 40 percent of the subjects yawned in response to the video. There was a tight correlation between a high score on the self-awareness/empathy test and the susceptibility to contagious yawning. The scientists hypothesize that people who yawn contagiously are both more self-aware and more skilled at reading the thoughts of others by observing their faces. A follow-up fMRI study showed that viewing someone yawn evokes activity in parts of the brain involved in these skills. "Yawning may be more of a reflection of our nature as social beings than of our sleep cycles," says Platek.

So here's a new gauge of character and the potential for friendship: Yawn and see who yawns back.

What's going on in the body during this hiatus? By midafternoon, are we just tired, tuckered out from our half day of exertion? According to Carskadon, young children don't experience the midday trough, even after plenty of physical activity, but kids in mid- to late puberty do. During adolescence, the dip becomes entrenched in our days and is present nearly every afternoon of life thereafter. In the elderly, the window of weariness broadens to include the period between 11:30 A.M. to 5:30 P.M.

Fatigue does accrue over the day, depending on your level of activity during the morning (and, of course, on how much sleep you got the night before). Once, when I was in high school, I played Helen Keller's mother in a series of afternoon performances of *The Miracle Worker*. I remember peeking out from behind the curtain before one show and spotting my own mother in the third row of the auditorium. She sat up, head erect, looking straight at me, it seemed, but her eyes were closed. When I got home later that afternoon, she had left me a note: "I don't know how you play that role day after day."

I stepped into the shoes of the caretaking mother for three hours a day a few times a week. My mother lived that role twenty-four hours a day for well over a decade. The effort of doing so—of feeding, bathing, ministering to my handicapped sister while caring for a family of seven—stole virtually all of her stamina.

Even those of us who aren't saddled with this sort of extreme caretaking job are fatigued from working long hours—longer than those worked a generation ago. Since then, Americans have managed to squeeze the equivalent of at least another week's worth of labor into the work year. The problem is this: When one packs life with unrelenting activity and little chance to relax and refuel, the body begins to feel run-down. Some evolutionary biologists view this kind of fatigue as the body's smoke detector or warning signal, its way of telling us to slow down to prevent the physical and mental harm of overexertion.

A deep seasonal rhythm largely ignored by modern society may also contribute to fatigue. Lethargy is one of the chief symptoms of seasonal affective disorder (SAD), a reaction to the waning sunlight of winter. A Norwegian friend of mine once told me about a word used to describe the season in her country, *morketiden,* or murky time, when a drape of darkness falls not just on the winter landscape but on the inner world of the soul. Underlying SAD is the persistent rhythm of our circadian pacemaker, which still has the capacity to detect and react to seasonal changes in day length, says Russell Foster, a circadian biologist at the Imperial College Faculty of Medicine in London. In response to the shorter hours of daylight, the brain secretes melatonin for a longer period during the longer night hours, putting the body in "night mode." It also reduces its production of serotonin, a neurotransmitter involved in regulating mood. But in modern society we don't slow down to accommodate our shifting seasonal chemistry; in winter, we continue to work long hours and stay up late at night, and our bodies suffer. For a small percentage of people—greater in countries at high latitudes—the wintertime decrease in daylight and overproduction of melatonin may cause a full-blown case of SAD: weight gain, reduced physical activity, and overwhelming fatigue. Daily exposure to a light source for prescribed periods can ease symptoms of the disorder.

Despite mounds of research, science is still struggling to understand commonplace fatigue. It ranks as one of the most frequent health complaints in this country, accounting for up to fifteen million doctor visits a year. But despite its pervasiveness, it is not an easy state to quantify, or even define.

Fatigue is not just feeling sleepy. You can be sleepy without being fatigued, and fatigued without being sleepy. It can be physical—the sensation of weariness in the body; or emotional—feeling unmotivated and bored; or mental—lacking concentration or sharpness. It can be diminished by encouragement and motivational cues (researchers have found that the perception of fatigue can be manipulated during a physically challenging activity simply by providing feedback on performance), or by financial incentives (one study showed that people promised a $5 reward were able to hang from a horizontal bar for almost twice as long as control subjects or subjects who were merely encouraged by the experimenter), or by the suggestion that the effort one is expending is less than it was before—even if it's not.

Fatigue can be so severe that it's incapacitating and yet, in fear or excitement, quickly forgotten. It may arise from grief, disappointment, physical illness, pain, the malignant and exhausting gloom of depression, lack of sleep, or relentless labor. So stymied are some scientists by the elusive quality of the concept that they argue it should be dumped altogether.

I hold with my mother's definition: Fatigue is the enemy.

In any case, exhaustion from a half day of wakefulness is not the foe we face each afternoon. After all, our energy often picks up following the midday lull. Nor is the culprit an ebb in body temperature, as is the case with parallel dips in the very early morning hours. What is the offender then? Is the post-lunch slump a consequence of that turkey sandwich and corn salad devoured earlier in the warm sun of the veranda?

Some evidence suggests that a big meal may contribute to the lassitude. "Gastric stretch" is thought to have some sleep-inducing influence (just as the absence of food may have an arousing influence). So, too, the movement of food from the stomach to the duodenum may exacerbate drowsiness. In cats, the mere act of gently stimulating the lining of the small intestine brings about acute sleepiness. Insulin, too, may play a part. Immediately after a feast, the body often experiences a temporary boost in energy from the rise in glucose, or blood sugar. But then follows a surge of insulin, the hormone that transports sugar to cells. In an effort to store the excess of sugar, insulin may extract too much of it from the blood, leaving little free for immediate energy. A big meal rich in fat may worsen the nosedive in alertness and performance, say researchers from the University of Sheffield, possibly because fat triggers the release of CCK, that hormone of satiety, which has been shown to cause sedation in humans and other animals.

However, research shows that the dip occurs whether or not you eat lunch. When scientists compared midafternoon sleepiness in young men who ate a heavy lunch, a light one, or no lunch at all, they found that 92 percent of the eaters napped afterward for a solid ninety minutes, regardless of the size of their meal; the fasters slept too, though only for thirty minutes. Lunch may aggravate or prolong the doldrums, say the researchers, but it doesn't induce them.

No one is sure what precipitates the afternoon trough. The work of Carskadon and others suggests that it may arise from a glitch in the timing of two opposing processes at work in our lives. First there's the homeostatic sleep mechanism, which acts like a sleep thermostat, keeping track of how long we've been awake. The need-to-sleep tab starts running as soon as we get up in the morning and tallies our sleep debt over the hours. As it builds, this homeostatic force exerts more and more pressure to discharge the debt, and we grow sleepier over the course of a day.

Every hour and a half to two hours, we feel an especially strong wave of sleepiness. Peretz Lavie confirmed this when he tested people's propensity for sleep by asking them to try to fall asleep every 20 minutes over a 24-hour period, a total of 72 attempts. Lavie found that a "sleep gate"—a window of "sleepability" when we may drift off relatively easily—swings open roughly every 90 to 120 minutes. The cycle is most pronounced at night (when shift workers may suffer alternating bouts of clarity and intense sleepiness), but it's also at work during the day. This pattern of sleep vulnerability occurs whether or not you have slept well the night before.

What keeps us awake and alert through these periodic swings of the sleep gate is the other process shaping our days: the circadian alerting mechanism, controlled by our central pacemaker, the SCN. Dale Edgar of the Stanford University School of Medicine verified the location of the alerting mechanism in a study of squirrel monkeys. The species has human-like sleep-wake patterns—staying awake for about sixteen hours, then sleeping solidly for eight: When Edgar destroyed the monkeys' SCN, they fell asleep again and again all day long.

Over the course of a waking day, this circadian alerting system beats a different rhythm from the sleep thermostat, says Carskadon. The alertness signal is lowest in the very early morning, say around 3 A.M., when body temperature is at its nadir. As the day goes on, the wakefulness signal gets stronger and stronger, counteracting the growing homeostatic pressure for sleep. A potent wave of alertness sweeps over us a few hours after waking, which may account for our late-morning mental agility. By early evening, the alerting signal is so powerful that it creates a "wake zone" a few hours before it begins its downward slide into circadian night.

Throughout the day and night, then, your body is subject to the push-pull of these two processes. For most of the daylight hours, the alerting mechanism overrides the homeostatic drive for sleep and rouses our bodies into a bright-eyed state. But around midday, says Carskadon, "sleep pressure accumulates before clock-dependent alerting achieves adequate strength to offset sleepiness"; we're overcome with a surge of drowsiness, and the sleep gates open wide.

Just how severely you suffer the slings of this afternoon slump may depend on your chronotype, notes Carskadon. "Evening types tend to experience waves of greater amplitude," she explains, "with higher peaks of alertness and lower troughs of drowsiness, as well as a mountain of alertness in the evening." Morning types, on the other hand, "have a relatively flat alertness curve during the day, which then falls off rather dramatically in the evening."

But most people experience some midday dip or trough, says Carskadon. For those on the road, these are dangerous hours. Studies of fatigue-related accidents in Israel, Texas, and New York show that single-vehicle accidents (driving off the road, for example) are most common not only in the wee hours of the morning, between 1 and 4 A.M., but also in midafternoon, between 1 and 4 P.M. This double peak also appears in the temporal distribution of public-bus accidents in the Netherlands and railway accidents in Germany. In fact, studies from all over the world show an afternoon spike in sleep-related driving accidents. At around 4 P.M., drivers are three times more likely to fall asleep at the wheel than they are at 10 A.M. or 7 P.M. In a world where human error may cause accidents involving large numbers of people, these drops in efficiency can have a catastrophic impact.

Those of us listening to the sleep lectures in our comfy auditorium seats wouldn't have felt logy if we had been outside in the wild coastal weather of Amelia Island. The gusty breeze and roiling waves of an impending storm would have boosted heart rate, dilated pupils, slammed shut those sleep gates. But ordinarily we don't have the advantage of hurricane-force winds to keep us perky. So how else to rescue alertness?

There are two ways to go. Try to override the rhythm, bear down on your work or your driving or whatever task is at hand, and ignore the open sleep door at your own peril. Or briefly go through it: pull over at a rest stop on the highway or put your head on your desk or, if you're lucky enough to have a couch, stretch out and snatch forty winks.

"Very bad habit! Very bad habit!" Captain Giles says in Joseph Conrad's *The Shadow-Line,* chiding himself as he retires for an afternoon snooze.

Catnap, siesta, forty winks, rest involving sleep but not pajamas— a nap is technically defined as a daytime sleep episode of more than five minutes and less than four hours. Considered by many to be deviant behavior, napping has traditionally gotten a bad rap, disparaged as the unfortunate artifact of an overindulgent meal, stifling midday heat, or sheer laziness. One is "caught" napping—okay for kids, but for adults, a sign of weakness, sloth, or senility. Even the medical profession has traditionally viewed the tendency to nap with suspicion, indicative of poor sleep hygiene or disorders such as sleep apnea or narcolepsy.

I'm happy to report that in the past few years napping has achieved new status. Research shows that naps not only ensure a break at a time of day when we're definitely not at our best, they also have powerful recuperative effects on performance, out of all proportion to their duration.

Some wise souls have long suspected as much.

Napping is common in traditional cultures, from Papua New Guinea, where people favor a two-hour nap at noon, thus avoiding the scorching midday equatorial sun, to chilly Patagonia, where the Yahgan, when tired, will lie down for a nap anywhere and at any time, to the inhabitants of Pukapuka, an atoll in the Cook Islands, who differentiate more than thirty-five kinds of dozing based on soundness of sleep and the position and movement of the sleeper.

"You must sleep sometime between lunch and dinner," advised Winston Churchill, "and no half-way measures. Take off your clothes and get into bed." In World War II, the British prime minister managed to be alert and awake at all hours of the night. "I *had* to sleep during the day," Churchill said. "That was the only way I could cope with my responsibilities ... Don't think you will be doing less work because you sleep during the day ... You get two days in one—well, at least one and a half, I'm sure." President Lyndon Johnson, too, is said to have slipped on his pajamas at midday in order to sleep soundly for half an hour, which gave him the strength to work into the night.

Claudio Stampi, an Italian sleep researcher, has explored tales—some probably apocryphal—of famous personages who existed solely on sleep stolen during naps. Thomas Edison, for instance, was an incurable nighttime insomniac who worked incessantly to accrue patents, including the one he's chiefly known for, which badly aggravated his own condition. Instead of sleeping eight hours a night, which he considered a "deplorable regression to the primitive state of the caveman," Edison got by with frequent naps. Leonardo da Vinci is said to have slept for fifteen minutes or so every four hours, for a daily total of less than two hours' sleep. By so doing, he may have gained twenty years of bonus work time during his sixty-seven years of life.

It may not feel as if naps have a reviving effect; sleep inertia from a nap often leaves one feeling groggy. But we're generally not good judges of our own restedness. As the pioneering sleep researcher William Dement points out, study after study demonstrates that naps enhance alertness, mood, vigilance, and productivity in the later hours of the day, particularly for night-shift workers and for those forced to work for long periods.

According to Dement, important findings emerged when researchers at NASA tested the effects of napping on pilots flying long distances across the Pacific at night. During such long-haul flights, the reaction times of cockpit crews typically plummet, and pilots frequently fall into "microsleeps," brief episodes of sleep 3 to 10 seconds long. In the NASA study, some crews were directed to rest for 40 minutes during transoceanic flights, netting an average of about 26 minutes of sleep. A control group of cockpit staff on similar flights got no rest. On the no-rest flights, crews experienced a combined total of 120 microsleeps during the last one and a half hours of the flights, including 22 in the final half hour, when the plane was descending to land. The napping flight crews experienced only 34 microsleeps in the same period, and none in the last half hour. Their reaction time, vigilance, and alertness also improved.

"Everyone knows about the need for naps in transportation," says Fred Turek, a sleep researcher at Northwestern University and a speaker at the Amelia Island event. "But there's still very little being done about it." In his lecture, Turek showed two slides: one with the figure of $1 billion, the cost of a B-2 Spirit, the world's most expensive bomber; the other with $8.88, the cost of a Wal-Mart lawn chair used by B-2 pilots for power naps. Money well spent, says Turek, but not used often enough in this way.

Even for those of us with lives less strenuous than those of long-haul pilots and truckers, a well-timed catnap can improve alertness and mood. One study showed that for sleepy subjects taking monotonous early-afternoon drives in a car simulator for an hour or two, a midafternoon nap of less than fifteen minutes improved reaction time and reduced driving impairment as much as drinking two cups of coffee. Japanese researchers who recently conducted a two-week work-site study of factory laborers got similar results: a short nap on a reclining chair after lunch markedly enhanced job performance.

Naps can even augment perception. Sara Mednick and her colleagues at Harvard University tested subjects by presenting them with a visual perception task four times a day. The performance of those deprived of a nap deteriorated across the four test sessions. But those able to get a little shuteye between the second and third session—not just a rest, lying quietly with eyes closed, but a real nap, complete with slow-wave and REM sleep—showed substantial improvement in perceptual acuity.

In a later study, Mednick showed that naps also facilitate learning. Both nappers and non-nappers spent an hour in the morning learning to identify the orientation of three bars flashed on a computer screen. All of the subjects were then tested on what they had learned, first at 9 A.M. and again at 7 P.M. The napping group, which slept for an hour or more before repeating the test, outperformed the non-nappers in accuracy by 50 percent—but again, only if they slept deeply enough to have both REM and slow-wave sleep.

In early 2007 came news that a regular post-lunch nap may reduce the risk of coronary death. A study of more than twenty-three thousand Greek men and women, ages twenty to eighty-six, revealed that those who napped had a 34 percent lower risk of dying from heart disease.

In short, say sleep researchers, naps make you sharper, healthier, safer. Some businesses in Japan, Europe, and the United States are beginning to heed the research and are building naps into work schedules to boost safety and productivity.

So what is the best time to take a nap, and how long should it be? The latest siesta studies suggest that just fifteen to twenty minutes' rest sometime between 1 P.M. and 2:30 P.M. can relieve fatigue, boost cognitive performance, and recharge your mental batteries. Longer naps of, say, forty-five minutes to an hour may require some recovery time—about twenty minutes or so—while the grogginess of sleep inertia wears off. "Naps of this length, however, are sensitive to time of day," says Mednick. Morning naps will have more light sleep; late-afternoon naps, more "cleansing" deep sleep.

Taking a short sleep at midday is a natural response to our biological need for rest. The human body is "programmed" for a siesta, says Mary Carskadon. There should be no shame in taking one.

7

STRUNG OUT

●

INSTEAD OF GRABBING A NAP, perhaps you choose the chemical route to mental clarity, slipping out of the office and around the corner for a jumbo latte. Even before imbibing the black stuff, however, you've begun to feel less sleepy. Maybe it's the fresh air. Or maybe it's nerves.

The tension has been building over the course of the day, one hassle after another. Now, on the way back from the coffee shop, you're fretting about the work piling up on your desk, that snide comment made by your supervisor, your looming deadline, and the impossibility of making it home in time for your daughter's soccer game. As you step into the intersection, a horn blares, and you look up just in time to see a Ford Bronco barreling through the red light. You scuttle back to the curb, spilling your coffee, breathless, then angry, as you realize you've escaped getting creamed by inches. Your heart beats thick, your knees tremble. At this time of day, the tide of cortisol and other stress hormones in your body should be ebbing, on the way to their nighttime nadir. But suddenly they're surging into your bloodstream. If you weren't alert before, you are now, strung out and scared stiff.

William James once wrote that "the progress from brute to man is characterized by nothing so much as by the decrease in frequency of proper occasions for fear . . . In civilized life, in particular, it has at last become possible for large numbers of people to pass from the cradle to the grave without ever having had a pang of genuine fear." It may be true that we're a lot less likely than our forebears to face the horror of ending up as some other animal's lunch. But that threat has been replaced with other dangers, real and imagined. I think of diving under my desk in third grade during the A-bomb drills of the Cold War, of being in a car driven by a drunk teenage friend, of looking out the window of a prop plane and seeing one engine afire, of waking in the middle of the night to the sound of an intruder breaking my apartment window.

I think of an incident on a brilliant fall day, in retrospect more comic than scary, though it didn't feel so at the time. Late that afternoon, I was walking home from a friend's house with my young daughter Nell, both of us worn out from a long weekend of athletic activity. As we headed up the hill toward our house, we realized that something was setting the area's dogs to frenzied barking. Our neighborhood is an eclectic enclave of urban and rural houses at the edge of the city, where sidewalks are only a sporadic adornment to newer homes, where people have two, sometimes three dogs and still borrow a tractor to plow up the yard for corn and potatoes.

Out of the corner of my eye, I caught sight of a black mass obscuring the steps of the Victorian house across the street. Not ten yards from us was an enormous bull planted on the crabgrass of my neighbor's lawn, wide-eyed and stamping the ground with one hoof, agitated no doubt by the canine cacophony. He must have weighed a ton.

Goose bumps rose on my arms and neck. Nell looked at me: "Mom?" We both froze for an instant. The bull let out a bellow. I jumped, grabbed Nell's hand, and ran for our gate. I knew the bull was not bloodthirsty. I knew it was just a lost *Bos taurus* that had wandered into the neighborhood from the stockyards down the road. But still I had the panicky feeling of prey. My legs prickled, my knees shook. My hands trembled so I couldn't open the gate. Nell's nimble little fingers flicked it open, and we ran for the house.

From the safety of the front window, we watched the bull vanish around the back of my neighbor's house just as three police cars pulled up. A pratfall of heavyset policemen piled out of their vehicles and headed for the backyard. They disappeared for an instant. Then, suddenly, all five came scurrying back, eyes bulging, mouths agape, breathing heavily. "Buddy," one called to another, "I ain't never seen you move like that!"

For the next five hours the bull terrorized the neighborhood, thundering through flower beds and vegetable gardens, mowing down fences, even stumbling up the stairs of a front porch, before the police finally cornered him on the grassy swale between two houses, and a man from the stockyards shot him with a tranquilizer dart.

Runaway bull, reckless SUV, overbearing boss: the body's reaction is the same, a rapid-fire but sophisticated fight-or-flight response that affects nearly every aspect of our being.

It all begins with subconscious fear. "We're built to deal with danger first, and then to think about it," explains Joseph LeDoux, director of the Center for the Neuroscience of Fear and Anxiety in New York. In a series of brilliant studies, LeDoux has teased out the circuits in the brain that control fear and found what he views as two separate pathways to fright, which he calls the "low road" and the "high road."

The low road, says LeDoux, is the reason we're still here.

Fear originates deep in the brain, in that almond-shaped structure called the amygdala. When we see a possibly dangerous object or hear a threatening sound—say, the shadow of a predator or the hissing whoosh of a speeding car—a quick-and-dirty preconscious version of the stimulus, a small, rough-hewn, and fragmentary part of the sound or image, flashes along the low road. This ancient and primitive visual pathway doesn't travel by way of the "thinking" cortex but goes straight to the amygdala, without any conscious awareness and well before the complete image or sound is fully reconstructed in the mind. The amygdala sends an instant "look out!" signal that sets the body in motion to respond quickly to potential danger.

Meanwhile, says LeDoux, a more complete version of the stimulus also works its way along the high road, to the sensory cortex, where it's considered carefully, processed in detail, and analyzed to create a precise picture of the situation. The high road may confirm the danger. Or it may deem the fear inappropriate and the risk unreal—say, a large, dark tree stump rather than a hulking bull—and turn off the fear response.

But by this time, the amygdala has already triggered the body's defenses: the startle, the freeze, the hair standing on end, the mobilization for fight or flight. "It takes only twelve milliseconds to turn on the fear response in the amygdala," says LeDoux. "It takes three times that long, thirty to forty milliseconds, for the [same] stimuli to reach the sensory cortex." Those extra milliseconds can mean the difference between life and death; hence the evolutionary value of the low road.

Think back to that near miss with the Ford Bronco. The "look out!" message signaled by the amygdala was only the first step in your salvation. The warning message was picked up by the hypothalamus, at the base of your brain, which in turn sounded a chemical alert to your pituitary and adrenal glands, two small, bean-like glands situated atop your kidneys. These responded by releasing a flood of stress hormones that produced the familiar adrenaline rush, quickening your heart rate, stepping up your blood pressure, and supplying extra blood, oxygen, and fuel to your muscles, especially the large muscles of your legs. In the meantime, the bronchial tubes in your lungs dilated to bring in extra oxygen, which reached your brain to keep it vigilant and alert. The energy and fat stores in your body released some glucose and fatty acids to deliver more fuel. Those hormonal signals whistling through your body triggered the constriction of blood vessels supplying your skin and the release of the clotting factor fibrinogen to help stanch the loss of blood from any potential injury. The stress hormone cortisol set off changes in the immune system, readying it for damage to skin, muscle, and bone, and for possible infection. Your brain was meanwhile pumping out endorphins, which act as analgesics to reduce pain, if necessary. The quick surges of adrenaline and cortisol also served to boost mental acuity. At the same time, your body was slowing those functions it would not need in an emergency: digestion, reproduction, growth.

"The idea of this activity is to shift all resources to the parts of the body needed most to face an immediate challenge," explains Bruce McEwen, a neuroendocrinologist at Rockefeller University. This makes sense, of course: When we see a lion or an SUV coming our way, as McEwen points out, we're better off using our energy to scamper away rather than to digest an egg or grow a toenail.

In the past decade, scientists working to fathom the full nature of the fight-flight reaction, also known as the stress response, have turned up surprising news. "This kind of acute stress response is good for the body," says McEwen. "It's a protective reaction; it sharpens our senses, improves our memory, even enhances our immune response." In fact, stress itself is good, argues McEwen, as long as it's short-lived. Though a short bout of stress guzzles energy, it does wonders for performance and can actually deliver a sense of physical and mental well-being. The stress response is a brilliant system for facing short-term challenges, whether lifting a car off a trapped child, weathering a hurricane, giving a lecture, or outrunning a bull.

The emphasis, however, is on short.

What makes us feel stressed out is chronic, repeated, or excessive exposure to stress, the coping with noise, traffic, time pressures, and day-to-day worries about work and family, debt, aging parents, marital problems. "This kind of mounting psychological pressure, which causes us to lose sleep, stop exercising, and eat the wrong things, puts excessive wear and tear on our bodies," says McEwen. "This is the real danger." It's not unlike the difference between acute pain, which serves as an essential alarm system, and chronic pain, which has little purpose and does much harm. Chronic stress can put the body's response system into overdrive, overwhelm or derail it, so that it turns against itself, causing serious illness, even death.

Scientists have known for years that unremitting stress takes a toll on the body but have just begun to grasp how such long-term psychological pressure gets under the skin.

As the day at the office progresses and the wrinkles build—cranky boss, disorderly papers, family business shoved to the side—you find yourself jiggling your knee nervously and hunching your shoulders, perhaps ruing your decision not to buy a cookie with your coffee and wondering about reaching for that emergency chocolate bar in your desk. Your head begins to throb. You bend over to pick up a pile of memos from the office floor and feel a stab of lower back pain.

The word "stress" has been so overused that it has lost much of its meaning. It derives from the Latin *stringere,* meaning to draw tight. A Hungarian scientist, Hans Selye, was the first to coin the term, in the summer of 1936 in a short report to the journal *Nature,* entitled "A syndrome produced by diverse nocuous agents." Fourteen stressful years later, Selye published a thousand-page opus on the subject, which he generously dedicated to those in the world who suffer strain from "sustained wounds, loss of blood or exposure to extremes of temperature, hunger, fatigue, want of air, infections, poisons or deadly rays . . . who are under the exhausting nervous strain of pursuing their ideal—whatever it may be, to the martyrs who sacrifice themselves for others, as well as to those hounded by selfish ambition, fear, jealousy, and worst of all by hate." Which describes pretty much all of us at one time or another, including the author himself. Selye, who was known to work an average of ten to fourteen hours a day, seven days a week, also dedicated his work to his wife, who, he said, understood "that I cannot, and should not, be cured of my stress but merely taught to enjoy it."

In Selye's mind, stress was any kind of crisis in the body, any demand or damage, from starvation or sleep deprivation to a strenuous muscular workout, an infection, or a fear-provoking event. These days, scientists tend to define a stressor as something that disrupts the body's homeostasis, and the stress response as the myriad adaptations that eventually restore the balance. If the disruption is short-lived, the body usually recovers from it quickly.

But the stress response was not fashioned with the prospect of modern life in mind, says Bruce McEwen, the barrage of one stressor after another that seems to characterize contemporary existence. Our bodies have trouble telling the difference between an immediate, life-threatening danger and, say, perpetual quarreling with family or the ongoing grind of money worries. As McEwen suggests, a hand-to-hand struggle or quick dash to safety is not an appropriate response to an angry spouse or an insufficient paycheck. These kinds of common stresses can build up over the course of a day, a week, a year, and the stress response gets stuck in high gear until the mechanisms that are supposed to help us eventually begin to betray us. Moreover, we often make the wrong choices in response to such relentless stress, says McEwen. We eat more rich food, drink more alcohol, work harder, put in longer hours, stay up too late, stop exercising, and end up feeling even more even anxious, exhausted, ill.

A few weeks after the 9/11 terrorist attacks, I gave an afternoon lecture to medical students at the University of Virginia about recent discoveries in genetics and their effect on personal choices in medicine. It had been a difficult month. My nephew, a young man studying finance who was on his first trip to New York, had been in the south tower when it was struck. By some miracle he got out. So many did not. The days in the wake of the attacks brought waves of worry and grief. Like others, I struggled through the days and slept only fitfully, finding it difficult to concentrate and falling behind in my work. I was nervous about this talk, about taking on a subject with so much scope and complexity and not having time to prepare. To pull it together, I dispensed with exercise, decent eating, family time, and rest.

At the start of the lecture I felt confident, if a little warm and flushed, a sensation I wrote off as excitement and the effects of adrenaline. Slightly elevated body temperature is not a bad thing under normal circumstances, and indeed has been linked with improved performance (possibly, say some scientists, because raised body temperature enhances neurobehavioral function). But it's a matter of degree. When temperature exceeds about 101° F, mental and physical functions deteriorate. During the question-and-answer period after the talk, I began to feel woozy and confused, and I responded to the students' intelligent queries with vague, distinctly unintelligent responses.

"Does that answer your question?"

Not likely.

When I got home, I found I had a temperature of 103° F. It was the beginning of a bad case of pneumonia that put me in bed for a month.

As early as the 1900s, Sir William Osler noted that stress in the form of frenetic living may make us sick, but only recently has a rash of new studies offered concrete evidence for the way psychological stress may affect susceptibility to all sorts of diseases. Hormones are a key to the link. Excessive stress disrupts the normal circadian rise and fall of hormones, flattening the curves at their peak. A small, temporary dose of stress hormones is generally a good thing for the body, says McEwen; big, long-term doses are not. Constant high levels of adrenaline raise blood pressure, which can create scars or lesions in the blood vessels of the heart and brain where artery-clogging material called plaque builds up.

Too much cortisol can be equally deadly, causing bone loss and increased abdominal fat. The raised cortisol levels that come with chronic stress actually step up the rate at which our food is turned into fat and govern where that fat goes. McEwen and his colleague Elissa S. Epel, a researcher at the University of California at San Francisco (UCSF), found that the increased cortisol levels experienced by women under stress led to great accumulations of abdominal fat, even in otherwise slender women. The hormone turns on fat receptors in the abdomen and belly so that fat deposits build up there rather than on hips and buttocks—which heightens the risk for heart disease, diabetes, and stroke. Excessive cortisol also makes the liver produce more glucose for energy, McEwen explains. Normally the liver releases a lot of glucose at night. When that typical dose is boosted by cortisol, the excess circulates all night while we're at rest and not using it, setting our bodies on the path to weight gain, insulin resistance, and diabetes.

To make things worse, we often respond to long-term stress by seeking to self-medicate with high-fat food. In the days after 9/11, I shunned my usual healthy fare and dove with gusto into banana bread, pasta, chocolate chip cookies, anything full of fat or sugar. My friends, I soon learned, were doing the same. It's a common response to feeling strung out, seeking the calm produced by comfort food. "Stress makes us hungry," says McEwen.

Until lately, there was plenty of anecdotal evidence that eating that brownie or chocolate bar might temporarily soothe the nervous soul but little in the way of scientific confirmation. That changed with research from UCSF showing that the signals the body sends out after consuming high-calorie foods can temporarily ratchet back the activity of the stress hormone system. With long-term chronic stress, however, the excess calories can mean a hefty addition of extra fat—and an inability to bounce back from a stressful event. When scientists at the University of Maryland studied rats fed a high-fat diet for ten weeks, they discovered that the fat-fed rats recovered from stress much more slowly than rats kept on a normal diet.

Our immune system also suffers. Built to be sensitive to the signals of acute stress, immune cells carry receptors for cortisol and other stress hormones so they can respond quickly to injury and potential infection. A single stressful event tends to rev up the system, enhancing its performance. But unremitting stress has the reverse effect. Some 150 studies suggest that incessant stress dampens the immune response and makes people more susceptible to infection. People who endure stressful conditions for more than a month—debilitating grief after the death of a family member, divorce, or loss of a job—are far more likely to come down with a cold than those less frazzled. They're also apt to produce a weaker immune response to a vaccine.

So, too, cumulative stress slows the healing of wounds. One study found that in women caring for a relative with Alzheimer's disease—a highly demanding job that requires intensive, exhaustive attention—small skin wounds took an average of nine days longer to heal than wounds in control subjects. Psychological stress, it turns out, inhibits a key component of the early stages of wound repair, the secretion of chemicals involved in healing known as proinflammatory cytokines.

Even more insidious, perhaps, is the effect of constant stress and vigilance on learning, memory, and the very structure of the brain. The hippocampus, a part of the brain crucial to memory, has many receptors for cortisol and so is particularly vulnerable to hormonal excess. McEwen has shown that even temporary stress, if sufficiently severe, can shrivel the dendrites of hippocampal neurons, those long projections where the neurons receive signals. "This may be a protective mechanism against permanent damage," he says. "By withdrawing contact with other cells, the hippocampal cells may avoid being blown out permanently, like a circuit breaker." However, the hippocampus normally plays a part in shutting off the stress response by signaling the hypothalamus to stop producing stress hormones. So damage to the structure from long-term stress can result in the release of even more hormones, setting off a cycle of elevated hormones and further destruction.

A kind of reverse process goes on in the amygdala, with a similar nasty effect of raising anxiety and stress. Scientists have discovered that repeated stress makes neurons in the amygdala bloom with branching dendrites. This creates more low-road connections with sensory neurons and thus more routes of access for unconscious emotional information. As a result, even a small sensory component of a potentially fearful situation—an innocent detail taken in unconsciously—may trigger a burst of activity in the amygdala, heightening fear. As Joseph LeDoux sees it, this mechanism may account for certain types of "nameless" anxieties.

Perhaps most disturbing is news that stress frays our very DNA.

Look around your home or your office. Who looks drained and exhausted? Whose face sags? "People who are stressed over long periods tend to look haggard," says Elissa Epel of UCSF. We all know this intuitively. To figure out why it might be so, Epel and her team recruited fifty-eight mothers between the ages of twenty and fifty. Thirty-nine of the mothers were—like my own mother—primary caregivers for a child chronically ill with a disease such as cerebral palsy or autism. The women filled out a questionnaire to determine their perceived levels of stress. The team then studied the women's DNA, focusing on tiny structures called telomeres, which serve as timekeepers, informing the cell of its age. Telomeres cap the end of chromosomes like those little plastic caps on shoelaces, preventing the DNA within from unraveling. Every time a cell divides, a bit of its telomeres fray away.

After many divisions, the telomeres are so reduced that the cell stops dividing and dies.

In 2004, the team published its remarkable results: Mothers who suffered the most perceived psychological stress had substantially shorter telomeres. They also had much-reduced activity of an enzyme called telomerase, which helps to preserve telomeres. Compared with the cells of unstressed women of the same chronological age, the cells of the high-stress mothers had "aged" the equivalent of nine to seventeen years. The scientists suspect that the underlying cause of this premature aging may be an overabundance of chemicals called free radicals—triggered by stress hormones—which hinder the activity of telomerase.

How your body responds to stressful events may be determined in part by your genes. Look around again at your friends and colleagues. Who worries every detail? Who goes with the flow? While one friend may fuss and fret about this hassle and that worry, another seems unfazed by anything. Most of us know someone who never feels strung out or anxious, someone who seems to seek out stressful experiences and relish danger.

My friend Miriam is one such fearless spirit. Petite but powerful and ever optimistic, she appears to sail through life's traumas with ease and resilience. In college, her friends called her "Mitomim," after mitochondria, those minute powerhouses inside our cells. Miriam seems happiest when she's pushing body and mind, running up Mount Washington, say, or pressing forward on a flood of different work tasks. One year, not long after her forty-fifth birthday, she climbed the sheer face of the Matterhorn during the dark hours of predawn and in the midst of an ice storm—just for the fun of it. The next year she suffered a head-on automobile collision that nearly killed her; when she came to, she said, her first thought was whether to ruin her husband's morning by calling him with the news.

Then there's Mary, for whom the smallest dose of stress, mental or physical, is so toxic it tips her into despair. Mary falls into the category of people who can't adjust to such mild traumas as public speaking or confronting a boss. In response to the everyday stresses of life, she overexposes her body to stress hormones.

What accounts for the gulf between merry Miriam and brittle Mary?

Not long ago, this might have been considered a barren thread of inquiry, the disparity written off as a matter of something as nebulous as temperament or human nature. But lately there has emerged what some may consider a possible biological underpinning for these differences. Whether or not we are derailed by life's stresses is influenced by certain genes—specifically, by the length of our so-called serotonin transporter genes. These genes, which come in two sizes, short and long, affect the expression of the mood-regulating chemical serotonin. The genes made headlines a few years ago, when scientists at the National Institute of Mental Health showed that the short version was weakly linked to neuroticism, a tendency toward anxiety, self-consciousness, moodiness, and low self-esteem, which earned it the moniker "the Woody Allen gene." People blessed with two long copies (about 30 percent of the population) are more apt to respond with resilience to a run of stressful life experiences. Those with two short copies (about 20 percent of people) are two and a half times more likely to suffer seriously from stress. The rest, bestowed with one of each version, are moderately vulnerable.

This is not to reduce Miriam's buoyant character to a single snippet of DNA. Nor is it to say that possessing a double dose of the long genes confers complete immunity from the effects of chronic stress. No single gene bears full responsibility for tolerance of stress or tendency to high anxiety; hundreds of genes likely figure into the mix. But the new research does suggest this: The interaction of genetic makeup (two short variants of the gene) and experience (major life stresses such as disabling injury, trouble in an intimate relationship, long-term unemployment) may put Mary at the susceptible end of the scale, more than doubling her chance of succumbing to stress-related ailments. And two long versions of the gene may give Miriam a relatively slim chance of suffering, no matter how many stressful episodes come her way.

So what's a worrier to do?

Try to feel in control of your situation, says Esther Sternberg, an expert on the science of emotions and health and the director of the Integrative Neural Immune Program at the National Institutes of Health. Feeling that you're in the driver's seat, so to speak—able to take action to affect your own fate—banishes feelings of helplessness and panic and reduces the effects of long-term stress.

By way of example, Sternberg tells the story of a U.S. Navy pilot she once met who routinely flew F-14 fighter jets from aircraft carriers. The pilot admitted that when he flew under duress—say, while taking off or landing his craft on a carrier in the middle of the night during a rainstorm in the Sea of Japan—he felt all those things most of us feel when we're stressed. His heart raced and his palms sweated. But he was not completely stressed out, because he felt in control and able to use the mechanisms of the stress response to his advantage.

When you can't control your situation, says Sternberg, another strategy is to try to quiet your mind through deep breathing and meditation. Richard Davidson of the University of Wisconsin and his colleagues recently explored the physiological changes underlying so-called mindfulness meditation. In this technique, the meditator focuses on the moment, on the quiet awareness of breathing, allowing feelings and thoughts to wash over the mind without judgment or action. Studies have shown that mindfulness meditation can be a powerful antidote to anxiety disorders, chronic pain, and hypertension; one study even demonstrated that it had a significant effect on the rate that skin cleared up in patients with psoriasis.

Davidson's team studied forty-one employees of a biotechnology company. Twenty-five of the participants underwent an eight-week meditation program. When the scientists measured brain electrical activity before and after the eight-week session (and again four months later), they found that meditators showed more electrical activity than nonmeditators in the brain's left prefrontal cortex, a region previously shown to be associated with generally positive emotions—enthusiasm, optimism, confidence. Moreover, the team uncovered a direct link between this positive mental activity and health, specifically the robustness of the immune system. Meditators who demonstrated the boost in left prefrontal cortex activity also showed the most vigorous response to a flu vaccine; months after the experiment, they produced the highest levels of flu-fighting antibodies. The magnitude of the increase in left-brain activity predicted the magnitude of the antibody response to the vaccine—the greater the left-sided activation, the more antibodies produced.

Another remedy for nervous emotion is music. A short time after 9/11, in a church on a tree-lined street in my hometown, a chorus of hundreds gathered to sing Mozart's Requiem Mass in D Minor. The church was designed for a congregation of three hundred, but that day the crowd spilled over into the aisles and the vestibule and out the two massive front doors. Most of us were there because we sought comfort but could not bear the somber speeches, vigils, and formal ceremonies commemorating those lost in the attacks. Hearing neighbors and friends perform Mozart's final and unfinished masterpiece seemed a fit way to pay tribute to so many unfinished lives.

The mass for the dead began. I grasped only a little of the Latin — *requiem,* of course, and *recordare* and *lacrymosa*—so I listened instead for the pure, brilliant strains of music in which the smallest changes were meaningful, creating order, melody, and harmony out of chaos.

In *The Magic Mountain,* Thomas Mann's character calls music a "politically suspect" art because of the way it can move people by appealing directly to their emotions, swaying their moods, even inciting them to action against their better judgment. But by the same token, music has the power to remind, soothe, heal—which it did that day, in spades. Think of the way that music consoles at weddings, marches, funerals. Think of how some pieces send chills up the spine—the last movement of Beethoven's Symphony no. 5, for instance, or Samuel Barber's Adagio for Strings. Music with a quick tempo in a major key has been shown to produce in listeners many of the physical changes associated with joy: excitement, rapid heartbeat, release of endorphins, goose bumps. Music of slow tempo in a minor key elicits changes linked with sadness, an experience of "negative" emotion that, oddly enough, is considered rewarding by most people and sought after as pleasurable and comforting.

Not long ago, scientists at the Montreal Neurological Institute asked a group of musicians to choose music that evoked such powerful responses, and then took PET scans of their brains as they listened to their choices. The scans recorded intense activity along the neural pathways in the brain associated with keen pleasure and reward—the same pathways activated by eating, sex, and drugs. Other studies suggest that music can lower blood pressure and trigger the production of endorphins, those natural opiates released by the body in response to pain or stress.

It's interesting to note that other species share our susceptibility to soothing music. Dairy cows make more milk when listening to a classical piece such as Beethoven's *Pastoral* Symphony or a popular song like "Moon River" than when exposed to the rapid beat of Supergrass's "Pumping on Your Stereo" or Wonder Stuff's "Size of a Cow." The slower, more calming music appeared to reduce stress and relax the Holsteins, increasing their milk yield by a pint and a half per cow per day. I wish I'd thought to pump out some Puccini that afternoon to pacify our renegade bull before he stressed out the whole neighborhood.

Perhaps the oldest of all stress remedies are two I've always believed were best at undoing the damage of a day: humor and companionship. Now science is confirming my suspicions. People with strong social networks fare better in coping with stress, especially with respect to heart disease, immunity, and brain function, says Bruce McEwen. "Social support is a powerful talisman" against stressful pressures.

So is a good laugh. Allan Reiss and his colleagues at Stanford University used neuroimaging to peep inside the heads of volunteers and watch which brain regions grew active when they exposed the subjects to a series of forty-two cartoons deemed side-splitting by a cohort of similar age and background. The neuroimaging revealed that the comics roused not just the modern, thinking cortex used to analyze the jokes, but also the brain's ancient reward circuits, the mesolimbic regions—those same dopamine-rich areas triggered by alcohol and mind-altering drugs.

That humor sparks the brain's primeval salience and reward system suggests that laughter has been around for longer than we have and may have survival value. E. B. White once wrote, "Humor can be dissected, as a frog can, but the thing dies in the process and the innards are discouraging to any but the pure scientific mind." Teasing apart the neural roots of a good guffaw may seem an ideal way to unweave the rainbow. But I like knowing that laughter is stress therapy rooted in ancient neural threads of joy.

What most powerfully affects the stress equation, in McEwen's view, are the personal choices we make on a daily basis. For most of us, "the real problem is our modern lifestyle," he says, our habit of working too long and too hard, depriving ourselves of sleep, eating too much high-fat food, all of which feeds directly into our stress load and perturbs our normal stress response.

How to nudge the body in the right direction? Mellow out with friends, meditation, or music. Laugh. Most important, McEwen says, eat well, get enough rest, lay off the fatty foods and cigarettes, and, especially, get out and exercise—a great excuse to leave work a little early and hit the gym.

8

IN MOTION

•

T ASTE YOUR LEGS, SIR; put them to motion," cries Sir Toby Belch in *Twelfth Night*. Doing so will significantly reduce your anxiety, clear your head, even relieve depression. I try to jog after work a few times a week on the hilly streets of Charlottesville, which I prefer to any running track. (On a track or treadmill I feel as Robin Williams does, like a hamster.) The run keeps me in shape, but more important, it's the best cure I know for my own stress and malaise.

The word has been out for decades about "runner's high," but there wasn't much science to support the claim that exercise affects mood. That has changed of late. More than a hundred studies have found that aerobic activity reduces feelings of anxiety. People who work out daily feel the biggest benefit, but just fifteen minutes of activity two or three times a week can lift spirits for two to four hours after exercise.

Even a single brisk walk around a park offers relief from temporary anxieties such as stage fright. Not long ago, researchers put young musicians from the Royal College of Music in London to the test. When they asked each student to perform for them, they found that pre-performance anxiety stepped up the students' heart rate by about 15 percent. Then the team requested a second performance, but instructed half of the students to walk for twenty-five minutes before performing again; the other half watched a video. The walkers had significantly lower heart rates and reported feeling more relaxed and better able to concentrate on their playing than their sedentary counterparts.

The temporary feel-good effect of vigorous exercise was once attributed solely to endorphins. It's true that prolonged cardiovascular exercise—running, rowing, cycling—increases endorphin levels by as much as two- to fivefold. It's also true that a rise in endorphins is often associated with improved mood. But it remains unclear whether the two phenomena are linked: According to some neuroscientists, endorphins circulating in the blood don't easily cross the semipermeable blood-brain barrier to reach the brain. The uplifting effect may be due to a boost in levels of other chemicals, such as noradrenaline, serotonin, and dopamine—that chemical active in stimulating the brain's reward center. Most likely the mood lift comes from the interaction of all these chemicals and others, according to John Ratey, a professor of psychiatry at Harvard University. A bout of brisk exercise, says Ratey, is not unlike taking a bit of the attention-enhancing Ritalin and a bit of the antidepressant Prozac and putting them right where they need to go.

In fact, when it comes to relieving the symptoms of depression over the long term, regular moderate exercise may work as well as drug therapy. In a study called SMILE (Standard Medical Intervention and Long-term Exercise), James Blumenthal and a team of researchers at Duke University discovered that vigorous walking, jogging, or biking for thirty to forty-five minutes, three times a week, is at least as effective as powerful antidepressant drugs at lifting major depression and keeping symptoms at bay.

The team studied adults over the age of fifty who suffered from major depression. The subjects were divided into three groups: the first received medication; the second, a combination of medication and an exercise program; and the third, an exercise program only. After four months, subjects in all three groups reported fewer symptoms of depression. In follow-up studies over the next several months, the exercise group had lower relapse rates than the medication group.

Physical activity also offers relief for those with only mild to moderate depression. One study found that patients who worked out for a half hour three to five times a week said they had half as many depressive symptoms as they had before they began their exercise program.

It's hardly surprising, then, to find a flip side to this phenomenon. A longitudinal survey of more than 6,800 men and women showed that physical *inactivity* is associated with more depressive symptoms and lower feelings of emotional well-being.

How might exercise work its deeper mood magic? Some scientists speculate that improved aerobic fitness does the trick, or possibly a reduction in the amount of REM sleep one gets at night (which may be the way that some antidepressant drugs work, too). Blumenthal suspects that people who exercise may also feel more of a sense of mastery and self-confidence, the perception of being in control—doing something positive and health-promoting, he says—which translates into improved mood.

Even if you're not stressed out or depressed, there's another good reason to duck out of the office and head straight for the gym: Late afternoon and early evening are considered the optimal hours for many kinds of athletic activities. Your body is generally at its physical best late in the day. Your perception of exertion is low. Your muscles are most powerful and your joints most flexible. Your hands and your back are about 6 percent stronger than they are early in the day.

A late workout also better benefits muscle-building. Exercise in the evening, and you may gain as much as 20 percent more muscle strength than you would if you trained in the morning. At the end of the day you also literally breathe easier: Airways are most open late in the afternoon. Moreover, the heart works more efficiently, and reaction time is at its peak. This has to do in part with core body temperature, which typically rises across the day to peak in the late afternoon or early evening. For every 1° C rise in body temperature, heart rate increases by about 10 beats per minute, and the speed of nerve conduction quickens by 2.4 meters per second.

For all of these reasons, most sports records are set between 3 P.M. and 8 P.M. Swimmers swim faster then; runners run faster. For elite athletes, training and performing at these optimal times may bring some advantage. For the rest of us, exercise often just feels easier later in the day.

Still, if you're wedded to an early workout, don't despair. Studies suggest that morning trainers can reach higher work rates. Early in the day, when body temperature is relatively low, you may start at a lower work-rate level than afternoon trainers, but it will build gradually until your body temperature reaches its optimal level. By the end of the training session, you'll be working harder than late-day trainers.

Also, back pain is often less severe early in the day. Our erect posture subjects the disks of our backbone to pressure equal to several tons per square inch. This pressure can squeeze nerves leading from the spinal cord, causing back pain that tends to worsen late in the day, as gravity during our upright hours effectively compresses the space between the disks. By contrast, the spinal column lengthens at night, when we "unload" it by getting horizontal during sleep. As a result of this unloading and lengthening, body height peaks in the morning (by a fraction of an inch), and back pain eases.

Moreover, the early hours of the day are better for workouts involving balance, accuracy, and fine motor control. If evening better serves the runner and swimmer, morning aids the surgeon and the archer, and also the neophyte: Late morning is the best time to learn new motor skills and to remember complex coaching instructions.

Perhaps this explains a humiliating late-afternoon training session I once experienced with an expert archer. I should have known better.

"Face up, sight on your target, release." Allison Duck planted her feet on the narrow strip of grass behind the gymnasium, nocked an arrow on the bowstring of her powerful recurve bow, drew the string back to the point of highest tension, and smoothly let loose the arrow, which neatly pierced the blue ringing the bull's-eye.

She made it look effortless. Allison fell in love with archery in her native South Carolina when she was nine. Now six feet tall and powerfully built, with upper-body muscles that show the weightlifting she does to develop her draw strength, she has been shooting nearly every day for more than a decade and regularly gives lessons.

This was my first go at archery, so I studied Allison's example closely. Nearly any bodily movement—tying a shoelace, kneading dough, doing the Macarena—is best learned by watching the action of a model or teacher. The human brain is built for imitation. Even very young infants show some rudimentary imitative ability; within an hour or two, babies can mimic frowns, smiles, and other facial expressions. New research suggests that our brains possess imitative "mirror" systems, networks of special neurons that fire both when we perform an action and when we watch someone else performing the same action. When I see Allison draw her bow, my mirror neurons automatically simulate the action in my own mind. This helps me understand her motion and her intentions—what she plans to do next. These helpful neurons are found in many areas of the brain, including the premotor cortex and the rear part of the parietal lobe. During imitation, the network of mirror neurons in the premotor area is often more active during the viewing of the movement than the actual performance of it.

At this lesson, I did my best to replicate Allison's stance, positioning my feet so that my left side was facing the target and turning my face, chest, and hips slightly toward it, back straight. So far so good. But then a problem surfaced.

Archery has been described as a contest of the archer with himself. Calm immobility is vital. In fact, so essential is a steady hand that archers are said to try to release the arrow in the lull between heartbeats.

I'll admit here that stillness is not my strong suit. When I was training for the theater as a college student, one exercise always eluded me: the conscious effort to quiet every muscle in my body, in thighs, arms, neck, cheeks. I could not quell the tense, fidgety feeling. Certainly I failed when it came to relaxing my tongue. I hold with Pascal: "Our nature lies in movement; complete calm is death." After all, only 10 percent of the body's mass is meant for quiet contemplation and judgment; the remaining 90 percent is for action.

I nocked the arrow, trying, as Allison advised, to keep my shoulder locked in a down position, head upright, fingers relaxed on the grip of the bow. But my body began to twitch, and my arms to quiver. I yearned for the singular "catch" muscle of the clam, which can contract and then lock itself in that state, easing the clam's discomfort in the awkward posture of holding open its valves to catch prey.

No such molluscan luck. My muscles stiffened; my shoulders wobbled. "Fire away," urged Allison. At the moment of release, I swayed slightly to the right, sending the arrow straight into the wall of the gymnasium, where it lodged firmly in the corrugated metal.

"Well," Allison remarked wryly, "*that's* never happened before."

Would my balance and fine-motor control, my ability to imitate Allison and absorb her coaching instructions, have been substantially better at an earlier hour? I'll never know. In any case, I plan to stick with running. As one sixteenth-century physician wrote, "Every meuving is not an exercise, only that whiche is vehement." In my book, archery doesn't qualify. So what does? Some modern researchers argue that only sustained, vigorous activity such as running, swimming, or working out at a gym for an hour is sufficient to provide the full health benefits of exercise, especially in staving off heart disease. Others believe that more moderate activity once a day is enough to deliver at least some benefits and offers a more realistic goal for most people. The U.S. government recommends at least thirty minutes a day of moderate activity, preferably more.

But what's moderate? A slow run? An energetic walk? Pumping iron three times a week? How about vacuuming or waxing the car? Though common household and garden tasks would hardly seem to fall into the category of a vehement workout, some research suggests that such tasks may count as exercise, depending on how they're performed.

Not long ago, Australian scientists persuaded a dozen men and a dozen women to wear a head harness, nose clip, and respirator valve to measure the energy expended while they swept, mowed the lawn, cleaned windows, and vacuumed. It turned out that some of the more vigorous activities qualified as exercise of moderate intensity if performed for adequate duration and frequency—say, a half hour of brisk leaf raking or lawn mowing a few times a week.

Climbing stairs counts too, as scientists in Singapore discovered. The team enlisted more than a hundred men and women to ascend and descend twenty-two short flights of stairs (the average number in a typical Singaporean high-rise apartment building, where most of the city's residents live). The results were impressive. Descending the stairs was equivalent to an energetic walk at 2.6 miles per hour, and ascending was much like running at a pace of 6 miles per hour. Going up and down once expended almost 30 calories. Climbing stairs is an ideal activity for the masses, the team concluded: convenient, private, requiring no special equipment, and cheap.

Unfortunately, only a quarter of all American adults climb enough stairs or rake enough leaves or walk far enough to get even the bare minimum of recommended daily exercise, and close to a third are completely sedentary. Just 2.7 percent of us walk to work; less than half a percent ride a bike. Such a sluggish existence is a radical departure from our ancestors' aerobic way of life. Evidence suggests that early hunter-gatherers walked up to twelve miles a day and no doubt ran long distances as well.

Pascal was right: We are made for physical activity, not for sloth. Without a workout of some kind—walking, climbing steps, rowing, hunting—our bones thin and our muscles atrophy. Loss of muscle and bone from lack of exercise tends to start in our late thirties and early forties. By age fifty, the sedentary among us may have lost as much as 7 percent of our muscle mass; by age eighty, about 40 percent.

The good news is this: It's never too late to defy the decline.

Say you start your workout by lifting free weights. Stay with this routine and you'll witness the remarkable plasticity of muscle and bone.

As you lift, your arm muscles work in pairs, one raising the arm, the other lowering it by pulling the same bone in the opposite direction. As the one muscle contracts and exerts a force, its counterpart relaxes and stretches. The force of muscle against bone fosters the activity of bone-building cells called osteoblasts. The more powerful the pull of your muscle, the greater the stimulus for bone growth.

You may not be able to perceive your growing bone strength, but you will notice the transformation in your muscles. Resistance training works its muscle magic through two mechanisms: adaptations in the muscle fibers themselves and changes in the neural signals that fire them.

We're born with all the muscles we'll ever have, more than 650 of them. The proteins that make up muscle fibers are breaking down all the time, and new ones are being assembled. Whether your biceps grow, atrophy, or remain the same depends on the balance between the rate of buildup and demolition. To tip the balance, you must load the muscle. Hard training can double a muscle's size; even light effort can boost strength. In one study, subjects asked to grip an object with maximum force for only a second a day gained an average of 33 percent in grip strength within five weeks.

At rest, muscle fibers are soft and flabby. But every fiber is innervated by a neuron from the spinal cord. When stimulated by a nerve impulse, the fibers shorten, shifting from loose rubber to springy steel. Exercising builds muscle strength by reinforcing these nerve signals and synchronizing them.

This may help to explain why just *thinking* about an exercise can boost strength. Scientists have learned that putting people on a regimen of mental gymnastics—instructing them to think about bending a finger or an elbow, or flexing an arm muscle—actually strengthens the muscles involved in the action. A team of scientists at the Cleveland Clinic Foundation in Ohio asked a group of volunteers to perform "mental contractions" of finger and elbow for fifteen minutes a day, five days a week, for twelve weeks. The mental workout did not affect muscle size, the scientists found, but did significantly increase muscle strength, by 35 percent in the hand and 13 percent in the elbow—most likely because it served to reinforce the brain's nerve signals to the muscle. This may be the science behind the visualization technique that many athletes use to improve their performance, mentally rehearsing the motions of an athletic task before attempting it. However, the researchers emphasize that such mental exercise can never replace a regimen of daily vigorous exercise, including strength training.

To keep your muscle and bone mass throughout life, say the experts, you should load them by lifting weights two or three times a week. The new research on the gain from this kind of resistance exercise—making muscles exert near-maximal force for brief periods—is irrefutable.

There is a catch, though: Not everyone will benefit to the same degree. In 2005, a team from the University of Massachusetts, Amherst, looked at changes in the strength and size of the biceps muscles in 585 men and women after twelve weeks of pumping weights twice weekly. Men showed more increase in size compared with women, while women outpaced men in relative gains in strength. However, both men and women varied greatly in response to identical regimens of training. Some subjects gained little in muscle size or power, while others doubled their strength and increased muscle size by inches. At least some of the benefit from exercise, it seems, depends on your genes.

If you've never lifted weights before, or if you engage in a particularly challenging session, you may pay later. Muscle soreness generally peaks twenty-four to forty-eight hours after vigorous exercise. Stretching does not prevent it. My own worst case of this "delayed-onset muscle soreness," as sports scientists call it, came a few days after I climbed a volcano in Guatemala. The hike up the twelve-thousand-foot summit was slow and arduous, but it was the next morning's descent that undid my thighs. Later that week, I wandered in circles through the beautiful colonial town of Antigua, scrupulously avoiding street curbs. So painful was it for my quadriceps to execute the downward motion that I could negotiate a curb only by swinging my legs awkwardly out to the side to step down.

Muscle soreness results from so-called eccentric contractions—the lengthening of contracting muscles that occurs with downward motion (for example, lowering a weight or descending a steep slope). To teach this lesson to his students at the University of Aberdeen, exercise physiologist Henning Wackerhage asks them to step up onto a bench five hundred times with one leg and step down with the other. "Students are often convinced that the upward-pushing leg will be sore in the days after the exercise," says Wackerhage. "But they're in for a surprise. The upward-pushing leg is normally fine, whereas some of the muscles in the downward stepping leg are usually sore." Going uphill or hoisting a heavy load may seem like hard work (and it is for heart and lungs), but going downhill or lowering that load is harder on the muscles.

Delayed muscle soreness is caused by microtears in the muscle, which, after a day or two, become inflamed. White blood cells migrating to the site to help repair the tiny tears release chemicals that trigger pain—a protective device to signal injury and the need for rest.

On the bright side, muscles respond to the damage by growing back stronger and larger. Satellite cells scattered on the surface of the muscle fibers proliferate, migrate to the damaged area, and insert themselves into the muscle tissue. There they lend their protein-building resources to the muscle fibers. With their help, the fibers can pump out more proteins, thus not only repairing the tears, but going beyond to build themselves up. Muscles adapt with training so that they're more resistant to the damage of subsequent exercise and are repaired at a faster rate.

Perhaps you've shifted to running now. Our species is built for this in limb, lung, and heart, says the biological anthropologist Dan Lieberman: "We're capable of running at a wide variety of speeds and altering our breathing patterns to suit them, and we're able to make use of energy stored in our tendons and muscles."

When we run, Lieberman explains, we shift from the "inverted pendulum" mode of walking to a bouncy "pogo stick" mode, using the tendons and ligaments in our legs as elastic springs. Elasticity is that property that makes objects spring back to their original shape after being deformed. When your foot strikes the ground while running, your tendons and ligaments stretch, absorbing elastic energy from the impact, like a bow when bent; when your foot rebounds, the tendons and ligaments contract again, recoil, and release their energy. Through such stretching and recoiling, tendons do much of the work in running, relieving the labor for muscles.

The secret to swiftness, it turns out, is maximizing this bounce. Speed is not a function of how quickly you reposition your legs in the air, but how hard you push off the ground. Using a treadmill with a force plate to analyze runners of varying abilities at top speed, one Harvard research team learned that virtually all runners take the same time to reposition their legs, called swing time. But the fastest runners apply more vertical force to the ground with each stride, which results in greater upward force captured by those elastic tendons and ligaments. The difference between you and Marion Jones, then, is not your speed of limb movement but your "ground force," which determines how far you go with each bounce.

While we may be built for running, it's still strenuous and provides an aerobic workout, raising the amount of oxygen we extract from the air we breathe. How much aerobic activity we get in large part determines our fitness. Older people who engage in regular aerobic activity are more fit than their younger sedentary counterparts. With this kind of exercise, heart rate accelerates, shooting up to as much as triple its usual pace, and with it, the output—the so-called stroke volume of each beat. Blood also circulates faster, thanks in good measure to the tortuous architecture of the heart.

The importance of the heart's sinuous and loopy curvature was recently revealed through magnetic resonance imaging. British scientists showed that blood flows through the asymmetrically curved cavities of the heart with swirling movements that redirect the inflow and send it around, slingshot style, to the outlet of each cavity. When your heart rate steps up during exertion, the asymmetrical cavities pull vigorously back and forth, each helping to fill the other, and then sending the blood rocketing through the vessels, so that the average time it takes for a blood cell to travel the body's whole circuit is reduced from one minute to about fifteen seconds.

At the same time, your body shifts its priorities for where your blood goes. At rest, about 20 percent of the blood that leaves your heart goes to muscle, 24 percent to the digestive system, 19 percent to the kidneys, and about 34 percent to the brain and various other organs. But when you work strenuously—run, bike, swim—the amount going to muscle jumps to 88 percent, with the flow to stomach and kidneys reduced to a total of just 2 percent (which may help explain the potential for stomach cramps when you exercise hard after a meal).

Scientists have known for some time that aerobic activity has the net effect of making the heart pump more efficiently, reducing blood pressure and boosting blood volume and rate of flow. But only lately have they come to understand how this protects against heart problems. It turns out that the risk of heart attack is increased by inflammation, which can trigger the rupture of plaque and other events in the coronary arteries. Researchers have found that the drag of faster blood flow during exercise activates anti-inflammatory mechanisms in our blood vessels, potentially reducing the risk of both heart attack and stroke. Even lower-intensity exercise can diminish the chances of developing heart disease by raising the level in the bloodstream of "good" cholesterol and also by ridding the body of visceral fat cells, which release hormones that tend to inflame parts of the cardiovascular system.

If you want to get the most bang for your exercise buck, says the cardiologist Michael Miller, you should watch your favorite comedy while on the treadmill or swap jokes with your running partner: The benefit of laughter to blood vessel health is nearly equal to that of aerobic activity. In 2005, Miller used clips from the movie *Kingpin* to make twenty volunteers giggle and roar while he measured the dilation of their arteries and blood flow. The laughter provoked by the funny stuff seemed to make the endothelium—the protective inner lining of the blood vessels—dilate or expand, increasing blood flow by 22 percent. (Disturbing, stressful scenes from *Saving Private Ryan*, on the other hand, constricted arteries and reduced blood flow by 35 percent.) This suggests to Miller that laughter may be good for the heart, offsetting the negative impact of stress on the body's blood vessels. Miller does not recommend replacing jogging with jokes, but he does suggest a daily dose of fifteen minutes of hearty hilarity to supplement your aerobic exercise.

When do you stop your workout? When your thirty minutes are up? When you've finished your running route? When your body says "no more"? Most of us don't push ourselves long enough or hard enough to hit what endurance athletes call "the wall," that physical and psychological barrier that makes bikers wobble and runners buckle. Still, even the minor weariness we feel is real. Where does it originate, in the muscles or the mind?

On my longer training runs, I go about seven or eight miles before I get really tired. My friend Francesca Conte goes four times that distance. One of the best in the sport of ultramarathons, Francesca routinely runs races of fifty and one hundred miles, mostly on rough, single-track trails through the woods—and wins them. To do so, she trains intensely, running in the daytime and at night, in the winter, on rock trails slick with ice, and in the summer, when she sweats so profusely she loses as much as 7 or 8 percent of her body weight. Sometimes she runs so long and so hard, she says, that her brain shuts down and she can't navigate her route, calculate how far she has run, or even remember her own last name.

Trained as a scientist, Francesca is smart, conscientious, methodical—hardly a fanatic—but she tells stories about her own workouts that give one pause. Once, to keep ahead of a challenger late in a hundred-mile race, she ran seven consecutive seven-minute miles downhill, suffering extreme pain in her thighs with each stride. She won the race, but the next day her quadriceps were swollen and bruised purple, and she could hardly stand.

Another time, she set her mind on preparing for a big fall race by running the length of the Appalachian Trail in Great Smoky Mountains National Park, a seventy-one-mile stretch of rugged terrain. Despite the forecast for strong winds and heavy snow at higher elevations, she and four fellow runners drove all day to arrive at the trailhead at 7 P.M. and began running up the mountain in the dark.

Francesca loves to run at night. "It's like scuba diving," she says, "everything calm and quiet." This night was no exception, with a sky full of stars and the moon showing through the clouds. But ten miles into the climb, the wind rose, and the sleet started, followed by a torrential freezing rain. The trail was soon covered in thick, slippery slush, and Francesca's clothing was soaked through. "We couldn't stop for more than a few seconds because the wind would cause us to shake uncontrollably," she says. "This made it impossible to eat or drink. As time went by, I was getting weaker and colder, and it was getting harder and harder to move fast enough to stay warm." Tired, hungry, at serious risk for hypothermia, she pressed on. Ten hours later, she made it—but, she says, she feels she only "just survived."

Francesca and I have different definitions of tired. That she can push herself to such extremes and keep running well beyond the point of exhaustion begs the question of the nature of fatigue.

"It's the brain, not the body," Francesca told me. "The hardest stretches are the ones your mind is not prepared for, that you don't expect. When you hit bad weather, for instance, or when you're ten miles from the end of your hundred-mile race, and you see a steep hill you had forgotten was part of the course. Suddenly you feel completely exhausted. But it's not in your muscles; it's in your mind."

Science is beginning to back her up.

Hippocrates held that muscle fatigue from exercise resulted from a melting away of flesh. For the past century or so, physiologists have assumed that the sensation arises when muscles reach their physical limit—when they run out of oxygen or the body fuel known as glycogen, or when they produce excessive amounts of poisons, such as lactate.

However, certain puzzles have plagued this hypothesis. For one thing, fatigue is not always accompanied by a shortage of energy or oxygen. In fact, according to Timothy Noakes, an exercise physiologist at the University of Cape Town, South Africa—and an ultramarathon runner himself—muscles do not run out of anything during exercise. They do not use all their fuel stores, and they rely on only about 30 percent of their fibers for even the most demanding tasks. "There's no evidence that we use all of the work capacity of our skeletal muscles, even when we exercise to exhaustion," says Noakes. Moreover, athletes such as Francesca often seem to have a little something extra left at the end of a race, which allows them to pick up their speed—to run those final seven-minute miles, for instance. If muscles were somehow depleted or poisoned by their own byproducts, how would you account for this ability to rev up in the last miles of a race?

"No study has yet clearly established a direct relation between any single physiological variable and the perception of effort or fatigue," says Noakes. Like Francesca, Noakes believes that fatigue begins in the brain. To demonstrate the mental component of exhaustion, he and his colleagues put sixteen well-trained runners on treadmills and periodically asked them to rate their own perceived fatigue. At the start of the experiment, Noakes's team told the athletes they were going to run at full speed for ten minutes, when in fact they would have to run for twenty minutes. Between the tenth and eleventh minutes, when the runners were informed that they would have to run an additional ten minutes, their reported feelings of fatigue skyrocketed.

According to Noakes's theory, the brain has a kind of "central governor," which sets the level of perceived fatigue based on the expectations associated with a task and establishes a subconscious pacing strategy to shield the body from exhaustion and damage. It does so by monitoring a blend of cues. These include physiological signals from the muscles about their working rate and stores of energy and oxygen, as well as signals from the brain's center for temperature regulation. In Noakes's view, the central governor is what made those runners feel tired between the tenth and eleventh minutes, before it adjusted to the new information. When the brain senses that the body is reaching its limits, says Noakes, it responds through feedback loops to the muscles, triggering the sensations of fatigue. It uses conscious cues, as well, to set a pacing strategy, postponing fatigue until the expected end of a race and creating the sensation of overwhelming exhaustion only when it's time to quit. In this way, the brain protects itself and the rest of the body from catastrophic collapse.

Just what sort of signals serve the central governor, flashing between brain and muscles to regulate fatigue, remains largely a mystery. One possibility is a molecule called interleukin-6 (IL-6). After protracted exercise, blood levels of the molecule jump from sixty to one hundred times the normal. Giving IL-6 to trained male runners makes them feel tired, slows them down, and impairs their performance. It may be that endurance athletes like Francesca have IL-6 receptors that are less sensitive than yours or mine, say some scientists, so fatigue, for them, really is a different beast.

Noakes's theory is still controversial, but I like it for its elegant explanation of common experiences: the weariness Francesca feels at the prospect of an unexpected incline late in one of her races. Or the reverse: the feeling experienced by many of us amateurs that the first mile of a ten-mile race is somehow easier than the first mile of a four-miler, though there is no objective difference. In the longer run, our central governor tells us not to feel tired yet; it's still too early in the race.

You've finished your workout. Consider its benefits. "Exercise invigorates and enlivens all the faculties," said John Adams. "It spreads a gladness and satisfaction over our mind and qualifies us for every sort of business, and every sort of pleasure."

It's true. Moderate exercise actually makes us feel *less* tired because it builds both strength and stamina. It boosts mood. It buttresses muscle and bone and improves cardiovascular health. A 2006 study showed that it reduces the incidence of colds in postmenopausal women, perhaps by increasing the number of white blood cells known as leukocytes, which fight infection. It elevates sensitivity to insulin, thereby diminishing the risk of type 2 diabetes. And it controls weight.

Exercise may help to reduce caloric intake by making some foods seem too sweet to take in large doses. In 2004, a team of Japanese researchers reported that in athletes, at least, a good workout heightened sensitivity to sweetness. But by far the most powerful impact of physical activity on weight comes from its effect on the body's energy balance. An hour of strenuous exercise can burn off about a quarter of a day's energy intake and also raise metabolic rate. Even after a workout, we tend to burn more calories than we did before, and the effect may last for hours. New studies show that this stepped-up metabolism arises in part from the boosted blood circulation and body temperature, as well as the body's efforts to replenish its oxygen stores and remove lactate.

The Amish people of Pennsylvania beautifully demonstrate this phenomenon. Though they eat quantities of calorie-rich foods—pies, cakes, eggs, ham—their rates of obesity are extremely low, less than one-seventh the U.S. average. The key to their leanness lies in their active lifestyle. Exercise physiologists have found that Amish men walk about nine miles a day; women, about seven. In addition, the men engage in some 10 hours a week of energetic farm work (the women, about 3.5 hours), and 43 hours a week of more moderate activity, such as gardening (women, 39 hours).

Anyone worrying about weight control might take a hint from this. Researchers have calculated that decreasing energy intake or increasing energy expenditure by only fifty to one hundred calories a day can offset weight gain in about 90 percent of people. For most of us, this extra hundred could be easily sheared away by a little of the Amish lifestyle: gardening for twenty minutes, walking a mile, bicycling for a quarter of an hour.

Lately has come news that exercise raises not just metabolism but brain power. Here's the exercise benefit that takes my breath away: Working out triggers changes in the brain that enhance learning and memory and protect against dementia.

Some years ago, the brain researcher Henriette van Praag gave a group of mice free access to a running wheel and withheld it from another group. She found that the mice that ran regularly learned new tasks faster than the ones that didn't—and their brains made more new cells. Mice that ran five kilometers (3.1 miles) a day learned to navigate a water maze more quickly than their sedentary colleagues. When van Praag and her team examined the brains of the mice, they discovered that the runners made two and a half times more new cells in the hippocampus, that part of the brain central to learning and memory.

What might have triggered the bloom of new brain cells? The researchers found that exercise spurs the growth of capillaries around the brain, which increases blood flow, raises oxygen levels, and boosts amounts of brain-derived neurotrophic factor, or BDNF, a molecule so important in helping brain cells grow and thrive that the neuroscientist Carl Cotman calls it "the brain's wonder drug." The brain cells of the running mice also showed evidence of more synaptic plasticity, that mechanism vital to learning and memory.

"It's reasonable to speculate that the same brain changes observed in rats and mice in response to exercise may also underlie some of the improvements we see in cognitive processes in adult humans," says Art Kramer, a psychologist at the University of Illinois and an expert on the mental benefits of physical fitness. Indeed, new evidence suggests that exercise not only sharpens people's thinking, it can moderate—even halt—the cognitive decline that often comes with age.

This really is excellent news, especially given that so much new research points to the ways in which age sabotages the brain. Not long ago, Naftali Raz of Wayne State University and his colleagues measured five-year changes in the volume of particular brain regions in healthy adults. The team found widespread shrinkage, though it varied across regions. Substantial diminishment occurred in the cerebellum—that "little brain" behind the brain stem that manages movement, balance, and posture—as well as in the hippocampus, crucial for memory.

It's not clear how these changes in the size of brain regions relate to the deterioration of cognitive function that comes with older age. But the failings are all too apparent. As life slopes down from the twenties, so goes working memory, perceptual speed, quick processing of new information, the ability to resist distractions. As we grow older, we have more difficulty picking up new skills and more trouble comprehending texts, finding the right word (the tip-of-the-tongue phenomenon), and remembering the names of friends and acquaintances. This is not senility or dementia, but normal cognitive aging. Even my father, sharp as ever in his mid-seventies, admits he needs a bumper sticker reading "I brake for names."

Tim Salthouse of the University of Virginia believes that the gradual falling off of mental performance may be due to the brain's slowed processing of lower-level stimuli. If the mind is slogging slowly through basic incoming information, it has less time to spend on more complicated thinking tasks, such as the ability to plan, make decisions, multitask, update information, block out the din of the irrelevant, and sift through memory. "But just why mental processing slows with increasing age isn't clear," says Salthouse. "Maybe there's a loss of neurons, which results in more neural detours, or maybe there are age-related reductions in the quantity of neurotransmitters or a degeneration of myelin, the sheathing around neurons that's involved in communication between them."

Fortunately, exercise offers hope. One big Canadian study showed that physical activity over a lifetime is linked with lower risk of cognitive impairment and dementia of any type. The link was especially strong for women. This was confirmed in 2004, when researchers at Harvard studied exercise patterns and mental performance in eighteen thousand older women who are part of the Nurses' Health Study at the Harvard School of Public Health. Women who walked or engaged in other regular exercise did better on memory and other cognitive tests than women who were less active; in fact, said the team, the exercising women performed as if they were three years younger. Physical activity in middle age may be critical to this protective effect. Studies show that older people who exercised at least twice a week when they were middle-aged were 50 to 60 percent less likely than their sedentary counterparts to develop dementia or memory loss. The exercise seemed to benefit particularly those who carried the gene associated with a bigger risk of Alzheimer's in old age.

Art Kramer and his colleagues recently investigated the changes that occur in the human brain with exercise. Their study showed that physically fit subjects had less age-related shrinkage of brain tissue in areas critical to memory and learning than did less active subjects, and that fit seniors had more intense blood flow in frontal brain areas that are normally associated with attention in younger brains. Previous imaging studies have revealed that the young use these frontal brain regions to complete a variety of cognitive tasks. As we age, our brains show less specificity in carrying out these same tasks—possibly because we're recruiting new brain regions to compensate for losses in the efficiency of our neurons in these areas. "Perhaps cardiovascular exercise, by boosting blood flow, helps our brains turn back the clock, biologically speaking," says Kramer, thus restoring the efficiency of those areas we rely on when we're young.

Think of it as you're cooling off after a hard workout: The mind that initiates your swim, your run, your vigorous row, is itself altered, enhanced, protected by the rush of blood and changed body chemistry it originally set in motion.

Evening

If you can get through the twilight,
you'll live through the night.

DOROTHY PARKER

9

PARTY FACE

D USK, THE HOUR between dog and wolf. You're finally home from work, well exercised or not, and perhaps still feeling a little stressed. To leave behind the cares of the day, you may try the route to forgetting suggested by Shakespeare's Julius Caesar: "Give me a bowl of wine. In this I bury all unkindness." The end of the workday is as good a time as any to have a drink. Tolerance of alcohol peaks now, just in time for the cocktail hour. Time of day influences how quickly alcohol is metabolized and how much it affects different target organs and body functions. Alcohol ingested early in the day is more intoxicating than the same dose at twilight. In one study of twenty men, those who received a large dose of vodka at 9 A.M. performed worse on tests of reaction time and psychological functioning than did those who received the same dose at 6 P.M.

Steal a moment before you head off to an office party and sit with your wine or gin to watch the dwindling light. I love this time of day: the gloaming, crow time, transit into night, when clarity of form dissolves, when everything near becomes distant and blurred with failing light. The body delights in thresholds, the poet Theodore Roethke tells us. It relishes the sweet coming-out of sleep or falling into it and this thing we feel at the cusp of evening. The ruddy sun limning the horizon helps to dissolve tensions and slow the hour.

This time of day may in fact affect your perception of the ticking minutes. In the late afternoon and early evening, when body temperature peaks, the passage of time seems to slow a little. To the interval timer in the brain, a minute in real time may feel several seconds longer.

Drugs such as marijuana and hashish have a similar time-expanding effect. William James wrote about the "curious increase" in felt time that comes with hashish intoxication. "We utter a sentence, and ere the end is reached the beginning seems already to date from indefinitely long ago. We enter a short street, and it is as if we should never get to the end of it." That glass of wine or gin, on the other hand, may make time fly. Alcohol reduces felt time compared with clock time, possibly by preventing the brain from receiving as many sensory inputs per second.

Whether booze buries any unkindness, however, is a topic of much debate. Depending on the particulars, on person and situation, alcohol can either lessen stress or intensify it. A key factor may once again be timing. If you consume alcohol before the onset of a stressful event, the drink may reduce its impact, says Michael Sayette of the University of Pittsburgh—for the very good reason that the booze prevents you from fully experiencing the event. It's called alcohol myopia. Intoxication disrupts the brain's ability to appraise new information and to link it with stressful associations. In other words, a traumatic event after a cocktail may feel less nerve-racking because the drink ensures that you don't quite know what hit you.

Alcohol myopia can also relieve anxiety and depression *after* a stressful occurrence, says Sayette, as long as the alcohol is paired with some kind of distraction, such as a party. The combination quite literally keeps your mind off your worries. Without such distraction, drinking alcohol after the fact can have the reverse effect, exacerbating stress—dubbed by one investigator the crying-in-your-beer effect.

So much depends on dose. That first drink makes us cheery and talkative, perhaps a little unsteady on our feet; the second or third slurs our speech, disrupts perception, affects our body sway, and impedes our ability to notice our own mistakes. It all comes down to our blood alcohol concentration, or BAC, the ratio of alcohol to blood in the body. This, in turn, is governed by how fast we drink, as well as how rapidly the alcohol is absorbed into the bloodstream and the rate at which the body distributes and metabolizes it.

There are cold-potato formulas. BAC is usually expressed as a percentage, reflecting grams of alcohol per deciliter of blood. (For example, .08 percent is equivalent to .08 grams per deciliter, a ratio that would make most of us feel pretty drunk.) After a person starts drinking, the time it takes to reach peak BAC can range anywhere from ten to ninety minutes. An hour after consuming two beers on an empty stomach, a 160-pound man may reach a BAC of about .04 percent.

I can enjoy a couple of drinks in an evening. But beyond this, my body cries out, "No more!" My low tolerance is fairly typical of my sex. Women reach higher peak blood alcohol levels than men after consuming equal doses of alcohol and become intoxicated with less drink. It was once thought that this gender difference was a simple matter of size or body weight: Women, generally smaller in stature than men, reach a higher BAC with less alcohol because there is only so far for the stuff to go. In a big body, the liquor travels farther, grows dilute, and loses its potency—certainly in a man as big as, say, two hundred pounds. But according to work by scientists at Stanford University, the real divergence results from the makeup of body mass in men and women, and perhaps some gender differences in chemistry. Women have proportionally more body fat and less water than men of the same body weight. Because alcohol is dispersed in the body's water, women—with their lower volume of water—reach higher alcohol levels than men after consuming equal amounts. Also, women may break down and eliminate alcohol and its byproducts less efficiently.

But BAC is influenced by a host of factors beyond gender and body makeup: whether your stomach is empty or full, for instance (a full stomach slows absorption); and how much sleep you've had (in the sleep-deprived, alcohol hits hard, so that one drink may seem like two).

The moderate level of drinking recommended by most experts means one standard drink (a beer or glass of wine) a day for women, two for men. To keep dog from morphing into wolf, wrote the poet George Herbert, "drink not the third glass, which thou canst not tame when once it is within thee."

It's early evening. You've arrived at your party and begun to mingle in the crowded room, launching into a lively discussion with a colleague. Though you've had only a single glass of wine, your mind fails you when an acquaintance approaches: In the midst of introductions, you temporarily blank on his name. It's there, tantalizingly, on the tip of your tongue, but you can't for the life of you retrieve it, and you stand awkwardly for a moment before mumbling, "You two know each other?"

William James described this failure as a kind of intensely active gap in our minds: "A sort of wraith of the name is in it, beckoning us in a given direction, making us at moments tingle with the sense of our closeness and then letting us sink back without the longed-for term." It's one of the "seven sins of memory" described by Daniel Schacter, a psychologist at Harvard University. Schacter's research suggests that this particular active gap is rooted in the absence of meaning in most proper names, which he calls the baker/Baker effect. If I tell you I'm a baker, he says, I'm giving you information about what I do and how I spend my time, which helps build a house of memory. If I tell you my name is Baker, I'm just providing a meaningless term. This means that the memory is isolated, bereft of mental ties or links, and so is vulnerable to temporary forgetting.

Here's where some kind of association strategy may come in handy, linking a name with an animal or object. Or perhaps a technological solution to the dilemma, like the one devised by Hewlett-Packard: a special cell phone headset fitted with a small camera that focuses on your field of vision and connects via the cell phone to your personal computer, where it accesses a database of photos and names. Once it fixes on a face, it jogs your memory with a vocal prompt.

We often forget names; we rarely forget faces. Survey the sea of people at your party, and you can spot those you know in a fraction of a second. This gift for instantly recognizing familiar faces in different contexts, regardless of view, age, light, and pose, is an astonishing perceptual achievement. Machines generally fall short at the task. "Real world tests of automated face-recognition systems have not yielded encouraging results," writes Pawan Sinha of the Massachusetts Institute of Technology. By way of example, he cites one test of face-recognition software designed to identify passengers with terrorist links. The system had a success rate of less than 50 percent, and some fifty false alarms for every five thousand passengers.

The philosopher Ludwig Wittgenstein called the face "the soul of the body," and Shakespeare called it "a book where men may read strange matters." No, wrote Milan Kundera, the face is only an "accidental and unrepeatable combination of features. It reflects neither character nor soul, nor what we call the self." In any case, faces are a currency of social exchange, and the ability to recognize them is a

vital skill. "This man" becomes "my friend" or "my husband." We all
have momentary failures—when we stand in blank unrecognition as
someone greets us at a party as if we were old friends. For most of us
these are mere temporary lapses. But there are some who suffer permanent face forgetfulness.

My sister's friend Heather Sellers, a professor of English at Hope
College and a deeply gifted writer, can't recognize or recall the faces
of friends or even family members. Each time they reappear in her
life, they seem new and strange. Heather has severe prosopagnosia,
a bewildering syndrome only partly understood, which disrupts the
brain's ability to recognize and remember features of the human face.
"When I see a face, I assume I see exactly what you see," she told me.
"Faces aren't blurry, foggy, altered in any way. But what I remember
about them, what I keep—that's what's different."

Heather believes that she and other prosopagnosics have trouble
recalling faces because they generalize them in the same way that non-
foresters generalize trees and non–chicken experts generalize chickens;
they don't see and retain the details necessary to categorize them into
subtypes. "I can't describe lips, noses, face-bone structure, foreheads,
chins, even eyes," she says. "When I think of someone I know well, like
your sister," she told me, "I see her hair and feel her warmth, her energy. I see her in a beige linen blouse and can conjure gold earrings. I
know she has a face, but I can't tell you a single thing about it."

Like other sufferers of prosopagnosia, Heather uses alternative
strategies for recognizing people—nonfacial "handles" such as gait,
hairstyle, body outline, mannerisms, and tone of voice—but these often fail. "Winter is harder than summer," she says. "People are bundled
up, and their padding screws up their gait and their outline; sometimes all that shows is their face. In these situations, I can't recognize
even my closest friends." Not surprisingly, the prospect of casual social contact fills her with dread. "The worst situation is a party with
ten people I know fairly well," she says. "I know that I'm not going to
recognize them. So I'll be a nervous wreck and have to work incredibly hard to identify them and to hold it all together, to manage the
anxiety." She tries to avoid parties altogether, or if she can't, to take
with her what she calls a "seeing-eye human" who will whisper to her
the identity of friends and colleagues: "There's the provost, Jim. Coming in from the left is John S. from Psychology. There's Dede in the
brown dress and the bangles. Talking to us right now is Lynn."

Oddly enough, Heather didn't know she had prosopagnosia until she was forty. (If you don't know what it's like to recognize a face, she told me, you don't necessarily know that you aren't doing what others do.) Then she stumbled on descriptions of the disorder while researching schizophrenia for one of her fictional characters. When she read the accounts, she was astonished by how accurately they described her own experience. She signed up for a study at Harvard, where she was officially diagnosed in 2005. "I was relieved and elated," she says. "I felt I had a great excuse for all the horrible social encounters I've had. It was the best test I ever failed."

Some cases of prosopagnosia result from stroke or damage to a blueberry-size patch of cortex in the right brain just behind the ear, known as the fusiform gyrus. Imaging studies show that neural activity in this small patch surges when normal people view faces. People who have suffered lesions in the area neither recognize familiar faces nor remember new ones. Most cases of the disorder, however, are a mystery, perhaps rooted in subtle developmental or genetic problems affecting this and other brain regions. Surveys suggest that as many as 2 percent of people have some degree of face blindness.

"I've learned that facial recognition is an enormously complicated process," Heather told me, "involving not only the ability to 'read' the topography of faces, but also memory, sensation, and emotion. To me, it's not so weird that I can't read a face," she says. "It's stunning that you can."

For years scientists have argued about where facial recognition occurs in the brain and how it normally works. Do our brains have specialized face-recognition modules? In a recent experiment with monkeys, Doris Tsao of the University of Bremen found that 97 percent of the cells in the fusiform gyrus respond almost exclusively to faces —evidence that this brain region may be just such a module.

Do the millions of neurons involved in the process work together, orchestrating the myriad bits of information about shape of nose, size of eyes, symmetry of lips into a single familiar visage? Or do individual neurons have the ability to respond selectively to a given face?

The latter concept, known as the grandmother neuron theory, used to seem laughable to some: So you have a single cell devoted to Grandma? One to Hillary Clinton? And another to Mick Jagger? Indeed, the idea appeared to be pretty far-fetched until 2005, when a team of scientists, including Christof Koch and Itzhak Fried, a neurosurgeon at UCLA, showed that individual neurons are in fact surprisingly adept at "face-spotting." In a study of eight patients implanted with electrodes, the team found that a single neuron fired selectively in response to assorted pictures of the same celebrity. In one patient, the same neuron was triggered by seven different pictures of the actress Jennifer Aniston. "This neuron looks for all the world like a 'Jennifer Aniston' cell," one neuroscientist remarked. The researchers are quick to say that these are not literally grandmother neurons but cells that are wired to fire in reaction to something specific and familiar, such as a well-known face. Their response may be more memory-related than visual. "I suspect," Koch explains, "that if this patient were to lose these cells, he would still recognize Jennifer Aniston as a female face but might not know that it was the Jennifer Aniston who had that TV show and who used to be married to Brad Pitt."

How would single cells "encode" specific faces? Doris Tsao's work suggests that each face-recognizing neuron is "tuned" to a set of facial characteristics; each acts as its own set of "face-specific rulers," she says, "measuring faces along multiple distinct dimensions," such as size and shape of individual features—iris size, for instance, or distance between eyes. By combining the measurements of all these little rulers, Tsao proposes, individual face cells may accomplish the miraculous task of reconstructing a face in the brain.

In scoping out the party crowd, does one face catch your eye? The Tierra del Fuegans have an expression, *mamihlapinatapei,* which is listed in the *Guinness Book of Records* as the world's most succinct word. It refers to the act of "looking into each other's eyes, each hoping that the other will initiate what both want to do but neither chooses to commence."

What draws two people together? Scientists have found that both face and gaze send a profusion of visual signals about mutual interest, health, even good genes. Though we've been taught not to judge a book by its cover, Shakespeare was right: One reads in a face many strange matters—identity, expression, even intent. We all do it, probably hundreds of times a day.

Take gaze. We're alone among animals in possessing eyes that signal where we are looking. The whites of our eyes, which highlight the iris, allow us to make eye contact and tell us instantly the direction of someone's gaze. This enhances "gaze signaling," a key cue for communication and cooperative behavior. A team at University College London found that a direct gaze from an unfamiliar attractive face enhances its appeal and activates the dopamine circuits in our brain that are dedicated to predicting reward. By contrast, if that same face looks away from us, the activity in this area diminishes. The heightened dopamine activity is not rooted in the attractiveness of the gazer per se, but in the potential for interaction signaled by eye contact, *mamihlapinatapei*.

Whether a meaningful glance leads to some more intimate interaction depends in large part on snap judgments we make without knowing it. The sense of who we find attractive, says the latest research, may lie in heartless formulas for seeking healthy partners with good genes. We carry these formulas buried in our minds and respond to the signals that promise to fulfill them.

So, what are we looking for?

Facial symmetry for starters. Most of us prefer faces with neat bilateral symmetry, which may signal a strong immune system and the absence of genetic problems. (Asymmetries often arise during fetal development from biological stresses such as poor nutrition, disease, parasites, or inbreeding.)

The masculine or feminine quality of a face is another such beacon. A team of Scottish and Japanese scientists recently showed that both men and women are attracted to more feminized faces of the opposite sex. In the sculpting of our faces in utero and throughout life, testosterone helps to carve the more chiseled masculine facial features of men; estrogen helps to shape the softer, rounder features of women. The researchers manipulated photographs of faces by enhancing or diminishing differences between the sexes. Subjects rated as more honest and cooperative the male and female faces that had been feminized—rounded, with smaller jaws. Feminized male faces, in particular, seemed to convey to women a "good father" signal. The scientists speculate that this preference may actually have limited the extent of human sexual dimorphism in facial appearance.

News of reproductive status may also figure into facial attraction, at least for men. Craig Roberts and his team at the University of Newcastle reported that men find especially appealing the faces of women who are ovulating. It had long been thought that women didn't reveal when they were ovulating with any kind of visual signal. While most animal species advertise their fertility through ruddied rumps or splashy scent, we humans seemed to hide ours. But Roberts's studies hint at the captivating possibility that our faces are a giveaway. The team showed that men judged photographs of women's faces taken in their fertile phase to be more alluring than photos of the same women taken in the luteal, or non-ovulating, phase.

"This increase in facial attractiveness is subtle," says Roberts. It involves variations in lip color and size, pupil dilation, and skin color and tone. But, he says, in evolutionary terms, even such understated effects can have a substantial impact on reproductive success by raising a woman's profile at a time in her cycle when the probability of conception is highest.

A direct gaze, a feminine, symmetrical face, full lips, and dilated pupils; throw in a smile (a potent signal that, if sufficiently broad, may be read accurately at a distance of a few hundred feet) and you have the sum of visual signals we may read in that face across the room or by our side.

But there's something else going on here. Well below our visual radar and beneath the screen of consciousness are other sorts of messages—chemical signals that convey far more than we ever imagined.

As you wander among the party guests, consider what you're taking in as social cues. It may seem all talk and visual clues. But a growing body of evidence suggests that in the matter of social evaluation and attraction, smell may be at least an equal partner. "In my lectures, I ask whether the ladies in the audience are turned on by the smell of certain men," says Mel Rosenberg of Tel Aviv University. "Invariably, I receive a positive response." To determine whether attractiveness of the opposite sex was influenced by smell, a team of British scientists asked thirty-two young women to rate male faces on aspects of attractiveness, then exposed them to a dab of male underarm sweat and asked them to rate the faces again. After the whiff, the women found the men significantly more appealing.

Though we have fewer olfactory receptors than an animal such as a mouse or a dog, which sniffs its way to food and sex, this doesn't mean we aren't swayed by the subtle powers of odor. As we now know, our olfactory system is exquisitely sensitive, capable of distinguishing tens of thousands of odorants in vanishingly small amounts. Women are better at the task than men, say scientists at the Monell Chemical Senses Center in Philadelphia—at least women of reproductive age.

This boost in sensitivity may result from female sex hormones that kick in at puberty, and likely serves to help women detect poisons in food while pregnant and to bond with children and mates.

We've also come to realize the impressive nature of our own odors. According to D. Michael Stoddart, a zoologist at the University of Tasmania, we humans are among the most highly scented of the apes. Odor glands abound in our face, scalp, upper lip, eyelids, ear canals, nipples, penis, scrotum, and pubis. But most of our normal, healthy body odor, a musky scent, issues from sebaceous and apocrine glands clustered in our armpits, or axillae, which start to function only at puberty. Apocrine glands secrete an oily substance that is odorless until the vast populations of microorganisms living in and around the underarm hair follicles and shafts break it down to produce musky-smelling compounds. (Another example of our microbial partners shaping our ways.) These molecules are wicked out into the world by underarm hair, says Charles Wysocki of the Monell Center. Removing the bacterial habitat and the smell "antennae" by shaving may result in a reduction of odor. But inevitably the axillary jungle returns, and with it the full aroma of those fragrant molecules—among them, fatty-acid compounds much like those that serve as sex signals in other animals.

It has long been suspected that axillary glands produce a scent attractive to the opposite sex. In his book *The Scented Ape,* Stoddart quotes folk stories of "a young man who would woo a peasant girl by placing his handkerchief in his axilla during a dance. When the young girl perspired, he chivalrously produced his handkerchief to wipe the sweat from her face. The power and allure of his axillary scent was such that she immediately succumbed to his wishes." In rural Austria, it was formerly a practice for girls to keep a slice of apple in their armpits during dances, writes Stoddart. At the end of the dance the girl would present the apple to the swain of her choice, who would—gallantly or readily—eat it.

Indeed, "one of the reasons that dancing is so appealing is that it's an opportunity for people to smell one another up close," adds Mel Rosenberg, who met his own wife on the dance floor.

Why would the underarm, of all places, serve as such an excellent source of sexually attractive scents? Possibly because of our erect stance: In daily life, scents originating from the sexual organs are not usually perceptible. Because humans walk upright, the underarm is the ideal odor-generating spot—"an area that often contains hair that can greatly increase the surface area for dispersal," says Wysocki, "warmed to aid in dispersion, and positioned nearly at the level of the nose of the recipient."

But there's a puzzle here. "If armpit odor is a turn-on, then why are we reviled by it?" Rosenberg asks. The answer, he suspects, may lie in the habits of modern civilization, which put us nose-to-pit with complete strangers in buses, elevators, waiting rooms, forcing on us their intimate odors.

Here's a new view of that party crowd. Issuing from the armpits of friends, colleagues, distant acquaintances, are waves of airborne chemicals that may be capable of affecting your perception, behavior, mood, even your libido and your choice of a mate. The word "pheromone" (from the Greek words *pherin,* to transfer, and *hormon,* to excite) was coined a half century ago to describe powerful chemical signals released and received by individuals of the same species. Mice, for instance, send vivid signals in body fluids such as urine, and even, according to one new report, in sex hormones secreted from the eyes. These invisible messengers may excite mating, block pregnancy, and accelerate puberty.

The idea that humans might participate in such invisible forms of communication has been greeted with a great deal of skepticism. But evidence is mounting to suggest that we almost certainly do. Among the first clues to the existence of human pheromones came in 1971, when Martha McClintock, now at the University of Chicago, published a paper showing that the menstrual cycles of female roommates in a Wellesley College dormitory tended to synchronize over time. Later it was found that the same effect could be achieved by merely depositing a little underarm sweat from donor women on the upper lips of recipient women.

Recently McClintock's team discovered that odors from breast-feeding women affect women who aren't lactating—influencing not only the length of their menstrual cycles but their libido. When exposed to the breastfeeding compounds, nonlactating women reported a 17 to 24 percent boost in sexual desire. The researchers suggest that pheromones of this sort may have evolved as a way of regulating fertility within groups of women—for example, by signaling one another that the environment was a good one for raising young.

As for the potency of male odors: Dab a little sweat from underarm pads worn by men under the nose of women volunteers, and perception, mood, and menstrual cycle may all fall under its aromatic influence. George Preti and his colleagues at the Monell Center exposed women to male underarm odors and then monitored both their mood and their blood levels of luteinizing hormone, which affects the length of menstrual cycles and the timing of ovulation. Normally the pituitary gland releases this hormone in pulses that increase in size and frequency at the approach of ovulation. The women subjected to male underarm secretions experienced acceleration in the onset of their next hormonal peak. They also reported feeling less tense and more relaxed when they had that sweat from males present on their upper lip.

What possible evolutionary reason could there be for this? Preti and his team speculate that early humans may have had relatively little time to spend in the company of their mates; the female reproductive system evolved so that it revved up the approach of ovulation when a woman caught a whiff of her man.

And here's news to lift the nostrils: A woman may reveal when she is ovulating not just by her facial features but by her smell. Scientists asked women to wear a T-shirt for three consecutive nights during ovulation and another T-shirt for three nights during the luteal phase of her cycle. The team found that men judged the odor of a woman's shirt worn during the fertile phase as more pleasant and sexy than the odor of a T-shirt worn during the luteal phase, even after the T-shirt had been kept at room temperature for a week.

Until lately, how we might sense such subtle pheromonal signals was a black box. Science had believed that mammals detected pheromones only with the help of a vomeronasal organ—a specialized olfactory system that does not function effectively in humans. But in 2003 the late Lawrence Katz, a neuroscientist at Duke University, overturned this view by reporting that neurons in the body's main olfactory system can detect pheromones. Since then, several studies have confirmed that we don't need a special organ at all to sense pheromonal vibes; our normal smell machinery may serve nicely to sniff out the volatile chemicals.

Just what other signals are you broadcasting as you buzz through the beehive of a social gathering? Nothing less than your deepest personal identity—and, possibly, your status as an acceptable genetic mate. We've known for decades that mice possess individual odor "signatures," which can be read in great detail by other mice and used to pick their partners. Now it seems the same is true for humans.

Each of us bears a chemical calling card that confers on us our own unique odor and reflects our subtle genetic differences. Moreover, this "odor print" signature can be detected by others. Women are especially adept at identifying the odor of relatives, their children, and their mates, says Mel Rosenberg; males to a lesser extent. The source of our singular odor is a key set of genes known as the major histocompatibility complex (MHC), which plays a big role in our ability to fight disease. These are the most diverse of all the body's genes, the better to deal with the multiplicity of bacteria, viruses, and other potentially harmful germs. Women tend to prefer the odor of men whose MHC genes differ from their own. One study found that women rated these odors as "pleasant" and the odors of men with genes similar to theirs as "less pleasant." By selecting partners with genes different from our own, say the researchers, we can avoid inbreeding or enhance our children's ability to fight disease.

To confound this tale, however, is a surprising new addendum: Women look for a fleck of their own father in their partner's MHC genes. Martha McClintock and colleagues found that in populations with plenty of genetic diversity, a woman prefers odors from males carrying MHC genes that match some of those she herself inherited from her father. Why? Perhaps women prefer a partner who shares some of their own vigorous immune genes rather than someone with completely unknown immune genes. Or they may be avoiding too much of a good thing. While diversity in these genes is generally considered beneficial, too much variation may make the immune system trigger-happy, increasing the risk of autoimmune disorders—the body turning against itself. In any case, the best choice seems to be a small number of matches. What's remarkable, as the scientists point out, is that women possess an olfactory system so exquisitely sensitive that it allows them to perceive these tiny genetic differences.

Such scientific revelations suggesting the possibility that attraction is influenced by a well-timed shot of pheromones or the keen detection of MHC genes may seem ultimate examples of the power of science to "clip an angel's wings, conquer all mysteries by rule and line," as Keats wrote. But I don't find them so. For me, these revelations have a way of enhancing the mystery. We think we make our choices willfully, consciously, through careful contemplation of the options; we think we know all that sways us. But in truth, part of what makes your marrow beat and your blood leap in wordless song may be a deep-down, chemical intuition aimed at protecting your unborn children.

Night

At night, every cat is a leopard.

ITALIAN PROVERB

10

BEWITCHED

PERHAPS YOU'RE HOME NOW, settled into your nest with partner or mate. Darkness has fallen, the hours of intimacy when
smell, hearing, and touch rule. Night has always yielded pleasures denied by daylight, offering privacy and refuge. As Shakespeare
wrote, "Light and lust are deadly enemies."

The hour or so after 11 P.M. is the most popular time for sex, but
not because of any intrinsic natural rhythm. When scientists studied
the circadian distribution of human sexual behavior, they found that
the majority of sexual encounters took place at bedtime solely because
of the rigidity of work schedules and family obligations. (Just what
qualifies as bedtime may be an issue for couples. Not surprisingly, research suggests that couples with incompatible chronotypes—larks
paired with owls—rate their marriages as less satisfactory than well-
matched couples, with more arguing, less time spent on shared activities, and less frequent sex.)

Unlike our mammalian relatives, who generally time their sexual
acts to maximize reproductive success, our cultural clocks and habits have corralled sexual behavior so that our preferred time for sex
is rooted in expediency rather than drive; it doesn't mesh particularly
well with our natural hormonal rhythms or fertility cycles. Levels of
testosterone, for instance, are significantly lower in the late evening
and higher in the morning, cresting at about 8 A.M. Semen quality, on
the other hand, peaks in the afternoon (with 35×10^6 more sperma-

tozoa per ejaculation than in the morning). This higher sperm concentration in seminal fluid probably derives not from circadian variations in sperm production and maturation, say researchers, but from variations in the nerve-muscle mechanisms that control ejaculation. Whatever the reasons, some experts advise that couples attempting to conceive have sex not at night but in the afternoon.

So much for the cold timing of making love. More than 1,500 years ago, Sappho described the physical symptoms of love itself: the blind eyes, "the elusive little flames that play over the skin and smolder under," the faintness and stupor. Since Sappho, we have not learned a great deal about the anatomy and physiology of love. Our understanding of such positive states as pleasure, happiness, and sexual arousal has not advanced nearly as spectacularly as our grasp of stress, anger, and fear. Perhaps such knowledge of love just isn't possible. "How on earth are you ever going to explain in terms of chemistry and physics so important a biological phenomenon as first love?" Albert Einstein wondered. But even the more straightforward aspects of sex are still cloaked in mystery—the brain mechanisms that control arousal, say, or the hows and whys of orgasm, which are awkward to pin down in a laboratory situation.

Of late, however, science has made valiant attempts to put under the microscope some of the more elusive aspects of love and sex, and in so doing, found glimmerings of their workings. Take the biology of a caress. Neuroscientists have lately stumbled across clues to the nature of our responses to this soft, stroking variety of contact in one who seemed to have lost touch.

Among the great physical pleasures of coupled life is the swapping of slow back rubs between partners, the methodical moving of hands along the swale of spine and up to the tight muscles of neck and shoulders; then "spoonerizing," as it's called in my household, curling into each other, comforted by the contact.

Unlike our other senses, touch is ubiquitous in the body, with receptors nearly everywhere—inside and out—that record sensations of pressure, pain, heat and cold, movement, and the awareness of where we are in space. Touch is the sense least easily fooled throughout life, the first one awakened in the developing fetus, the last to leave us at the end, and perhaps the most essential to our well-being.

Human infants deprived of touch fail to thrive. When scientists visited the overcrowded, squalid orphanages of Romania after the overthrow of the dictatorial Ceauşescu regime, they found that the hundreds of babies who were rarely or never touched were developmentally disabled and had high levels of cortisol. While there were many causes of their trauma, deficiency of touch seemed to play a key role in exacerbating their stress.

By contrast, ample touching, especially massage, has been shown to reduce levels of stress hormones and boost oxytocin, the hormone of pair-bonding and maternal love, which has a calming effect, lowering heart rate and blood pressure. The purported positive effects of massaging touch are legion, among them diminished pain, better lung function in asthma patients, even improved alertness and performance in children with attention disorders.

Touch is as old as life itself, going back to those early single-celled creatures that gained sensitivity to dimpling or pressure on their protective outer layer. In humans it arises from nerve endings beneath the surface of the skin that sense physical strain or pressure and convert this mechanical energy into electrical signals that travel to the brain. These nerve endings are distributed all over the body but cluster most densely in the lips, tongue, fingertips, nipples, penis, and clitoris. Among them, it appears, are some specialized to detect a caress.

Not long ago, the neurophysiologist Håkan Olausson and his colleagues in Sweden studied a fifty-four-year-old female patient who had lost sensation from her touch receptors. The patient could not detect pressure or tickling and denied having any touch sensibility on her body below the nose. Yet she could detect the faint sensation of light skin-to-skin touch and reported it as distinctly pleasant. The case suggests that our bodies possess a system of touch receptors separate from the nerves that detect pressure and vibration. These "slow-conducting" nerve endings lie beneath hairy skin and are specifically tuned to soft touch; when stimulated, they activate areas in the brain involved in sexual arousal and the processing of emotions. "This kind of receptor is abundant in animals but was long thought to have disappeared during the evolution of humans," says Olausson. "The finding that we still have a special touch system dedicated to processing emotional or social aspects of skin stimulation suggests the paramount importance of tactile stimulation for human well-being."

Why is it that my husband's hand just dimpling the skin of my back like a water strider keeps me blissfully in the present? It's not anyone's caress I savor. Grasping the mechanics of loving touch is one thing, but getting a handle on the nuts and bolts of love itself is quite another. Still, science is trying.

Italian researchers probing the hormonal changes that come with courtship found that both men and women in the first mad bloom of love have more cortisol pulsing through their blood, suggesting that the state is both stressful and arousing. Perhaps more noteworthy are studies showing that love-struck men have reduced levels of testosterone compared with controls; smitten women have raised levels. This could be a simple result of stepped-up sexual activity, say researchers, or it could be that this hormonal convergence somehow facilitates the courtship game. More testosterone would make women more sexually assertive; less of the hormone would make men less aggressive — good beginnings for the formation of a strong pair bond, notes the anthropologist Helen Fisher.

Fisher, for her part, has peered inside the head to see what brain systems might be activated when we're in the throes of lust, romantic involvement, and long-term attachment. She and her team at Rutgers University conducted brain scans on young adults who had just fallen head over heels in love, as measured by the so-called Passionate Love Scale. This lab standard (a sexual counterpart to the Stanford Sleepiness Scale) asks what the subject feels in the presence of a loved one — trembling, pounding heart, accelerated breathing, or excessive energy. It also inquires what percentage of waking hours is spent musing about the "love object" and then rates intensity of feeling, from tepid to madly in love.

The team chose those subjects deemed wildly, recklessly in love, then used fMRI to look at the brain circuitry activated when the lovesick subjects viewed photos of their beloved as compared with photos of mere acquaintances. The picture of the sweetheart, it turns out, ignited neurons in the dopamine-rich reward system of the brain, the caudate nucleus and ventral tegmental area — the very regions that light up during alcohol and drug use. The love-struck also showed raised levels of noradrenaline and low serotonin levels resembling those of people with obsessive-compulsive disorder.

That the neurochemistry of new love is entwined with the reward system of the brain is not so surprising from an evolutionary point of view. What's interesting is this: When Fisher and her colleagues compared the brain activity of the newly infatuated subjects with those involved in longer-term relationships, they found a difference.

In long-term lovers, the view of a loved one sparked lots of activity in the brain regions devoted to emotion. But for those newly in love, the images triggered little activity in these areas. This result confirmed earlier findings by researchers studying volunteers who professed to have recently fallen "truly, deeply and madly" in love. Those scientists were surprised by how small an area of the brain (just that dopamine-rich region) was activated when beholding the image of the loved one: "It is fascinating to reflect," they wrote, "that the face that launched a thousand ships should have done so through such a limited expanse of cortex."

The whistling, wing-beating fury of early love is more like a compulsion than an emotion, Fisher suggests, a motivational drive so powerful that it resembles the urge induced by addictive drugs, making the brain focus solely on the longing for and pursuit of reward—in this case, the love object.

As love progresses from lust to crush to commitment, different sorts of biology and brain chemistry kick in, posits Fisher. Simple lust, which motivates people to seek sex with a range of partners, engages androgen circuits. Romantic love, which guides the pursuit of preferred partners, is linked strongly with dopamine systems. And attachment, which ensures that individuals remain with their mates long enough to support a child and to parent well, is associated with a web of neurochemical networks involving two hormones: vasopressin, which boosts male attachment, and oxytocin, which seems to regulate all manner of positive social interactions, including trust. (One widely touted study in 2005 of Swiss students playing an investment game reported that oxytocin delivered as a nasal spray increased the students' willingness to trust one another.)

Of course nothing in human biology is so neat. These systems may operate independently or they may overlap, says Fisher, and their activity differs in men and women. The pattern of neuronal firing in your besotted brain does not necessarily match that of your lover's.

"A man falls in love through his eyes, a woman through her ears," wrote one British politician. Indeed, studies of sex differences in the processing of sexually arousing images show that men display extra firing in visual areas of the brain, in the amygdala and hypothalamus, and, as Fisher puts it, in regions "associated with penile turgidity." Women are more sexually aroused by romantic words and themes in films and stories than by images, she says. Women in love also generally display more activation in brain areas linked with attention and memory earlier in a relationship than do men; later on, they show more activity in regions associated with emotions.

When it comes to mental processing in men and women, the list of studies suggesting distinctions is growing: language processing, spatial skills, navigation, sense of smell. Functional MRI studies show that during reading, men and women use certain language areas of the brain differently. In navigating the physical world, men are better at mentally rotating maps and tend to think in terms of cardinal directions, while women excel at remembering landmarks and use relative directions. (This disparity emerges only at puberty; before then, girls and boys use the same navigational style, which suggests that steroid hormones may bring out the divergence.) We're only beginning to fathom the nature of these gender differences in brain activity, especially when it comes to sex, but the gulf seems real. So it's somewhat unexpected to learn that orgasm is processed in the brain in a similar way for both sexes — even if thinking or reading about it isn't.

Called at once the "supreme ecstasy" and "*la petite morte,*" the intense wave of pleasure known as orgasm has been the subject of abundant literature; nonetheless it remains an abiding mystery. In his classic studies on human sexuality, Alfred Kinsey described orgasm as an explosive release of built-up neuromuscular tension — so intense in some individuals that it may cause a man (or woman) to "throw his whole body into continuous and violent motion, arch his back, throw his hips, twist his head, thrust out his arms and legs, verbalize, moan, groan, or scream, in much the same way as a person who is suffering the extremes of torture."

We know that orgasm is the result of contractions in the pelvis and the perception of pleasure in the brain. But until lately, we haven't grasped how the two phenomena are linked.

In men, orgasm usually coincides with ejaculation, but the one can be experienced without the other. Erection, the necessary prelude to ejaculation and orgasm, often begins with tactile stimulation, especially of the glans penis, which sports a high density of tactile pressure receptors. The sensation of touch travels along sensory nerves to the lower spinal cord, which causes blood vessels in the penis to dilate and blood to rush through hundreds of corkscrew-like vessels to the organ's spongy tissues at fifty times the normal rate.

All of this can happen without conscious control. In fact, most of the erections experienced by young men, totaling around three hours over the course of a day, occur primarily during sleep. As Leonardo da Vinci wrote, with characteristic irreverence, the penis "sometimes displays an intelligence of its own; where a man may desire it to be stimulated it remains obstinate and follows its own course; and sometimes it moves on its own without permission or any thought by its owner. Whether one is awake or asleep, it does what it pleases; often the man is asleep and it is awake; often the man is awake and it is asleep; or the man would like it to be in action but it refuses; often it desires action and the man forbids it. That is why it seems that this creature often has a life and intelligence separate from that of the man."

Some years ago, scientists at Johns Hopkins University School of Medicine discovered a controlling factor in the complex physiology of erection: Contributing to the blood flow that initiates and sustains an erection is nitric oxide, the same gas formed during a lightning storm and so essential to the heavy breathing that accompanies exertion. In the penis, nitric oxide acts as a potent muscle relaxant on the smooth muscles that surround the blood vessel walls, allowing the vessels to dilate. An erotic thought or tactile stimulation brings about an initial surge of nitric oxide from nerve endings in the region, which triggers the erection; then the blood vessels release more of the gas to sustain it. Eventually, an enzyme kicks in to break down the nitric oxide, the arteries constrict, and the animation fades. The drug Viagra works by interfering with this breakdown enzyme, allowing the nitric oxide to hang around longer and maintain the pressure.

So, too, we've learned something about what controls ejaculation. It's no simple knee-jerk reflex, as once imagined, but rather the upshot of complex coordinated actions of the prostate, seminal vesicles, urethra, and pelvic floor muscles. What triggers it is still little understood. A study by Lique Coolen, a neuroscientist at the University of Western Ontario, has revealed that a small cluster of nerve cells buried in the spinal cord of the lower back may generate the action. Rats in which this so-called ejaculation generator is destroyed are able to find their mate, mount her, and achieve an erection, but can't ejaculate. Coolen suspects that the ejaculation generator serves as a kind of way station, processing sensory cues from the genitals and erotic perceptions from the brain. It then sends out signals that control the muscular spasms of ejaculation and also informs the brain of its occurrence.

Work by Coolen also suggests that the cells in the spinal ejaculation generator may form synapses with cells in the brain's ventral tegmental area—a pleasure region activated during orgasm.

As for women: Estrogen plays little part in arousal, despite the derivation of its name from the Greek *estrus,* or "intense desire." It does prepare the vagina for sex, elongating and widening the vaginal tract and triggering the cells lining the passage to secrete droplets of lubricating fluid. But it's a weak version of the "male" hormone testosterone, made in a woman's adrenal glands and ovaries, that heightens the sensitivity and responsiveness of touch receptors in her clitoris, labia, and nipples. From these receptors and a rich variety of specialized nerve endings in the genital area—clitoral shaft, glans, urethra, and the so-called G spot, a zone of acute sensitivity—come the sparks of arousal.

Yes, the G spot is real, at least according to Italian researchers. The spot is thought to reside a couple of inches inside the vagina, on the front wall behind the pubic bone. Buried in the flesh here are glands comparable to the male prostate gland. The Italian team reported that the same enzyme markers of nitric oxide activity found in the erectile tissue of penises also abound in most women in the G spot region. Gentle pressure on the spot raises pain thresholds by 40 percent and causes oxytocin levels to surge up to five times higher than normal. Some scientists speculate that this rush of oxytocin may explain sex's calming effect. In 2006, British researchers found that having sex before a stressful event such as public speaking lowers blood pressure, an effect that may last for as long as a week.

The nerves responsible for communicating stimulating sensation from the genital areas to the brain issue from the spinal column. But scientists at Rutgers University studying women with spinal cord injuries say they have discovered a novel sensory pathway outside the spinal cord that may also convey sensations from the vagina and cervix directly to the brain. This pathway travels via the vagus nerve, which wends its long way from the brain stem through organs in the neck, thorax, and abdomen ("vagus" means wanderer), bypassing the spinal cord altogether. Thanks to this pathway, say the researchers, women who have suffered "complete" spinal cord injuries may nonetheless experience orgasm.

Why some women reach orgasm during intercourse and others do not has been a persistent riddle. One new study points an intriguing finger at heredity. A team at St. Thomas' Hospital in London asked thousands of female twins about how often they achieved orgasm during intercourse. Most of the women said only infrequently; a small percentage reported always experiencing it; and an equally small percentage said they never reached it at all. By examining differences in results among identical and nonidentical twins, the team found a clear genetic influence accounting for 35 to 45 percent of the variation. The nature of this influence, however, is far from obvious. It could reside in anything from personality traits to the anatomy of sexual organs to levels of enzymes and circulating hormones.

Most of us are surprised to learn that orgasm actually takes place not in the genitals but in the brain. A case report in the *Lancet* entitled "Unwelcome orgasms" shed light on this odd phenomenon. A forty-four-year-old woman reported having recurrent episodes of orgasm unrelated to any sexual activity, once every couple of weeks. "They had no definite triggers," the doctors wrote, "and were neither particularly pleasurable nor satisfying because they were out of her control. On several occasions she experienced an episode while driving and had to stop the car." It turned out that the woman had a vascular abnormality in her right temporal lobe.

Orgasm is, in reality, a cerebral experience, as the neuroscientist Jean-Pierre Changeux once said, "and it is in the brain we must look for it."

Dutch scientists shocked the neuroscience community by doing just that. Gert Holstege and his colleagues at the University of Groningen in the Netherlands used a PET scanner to view the brain regions activated in men who were manually stimulated to orgasm by a spouse or lover. A year later, they did the same for women. The results showed that women's and men's brains display roughly the same widespread pattern of neural firing—about a 95 percent overlap. (The main difference was in a midbrain area called the periaqueductal gray; this region, which plays a role in modulating pain, fired up only in women.) Most of the activity occurred in the caudate nucleus and ventral tegmental areas of the brain, the same dopamine circuits triggered by romantic love and by drug use. In fact, brain activation during orgasm closely resembles the pattern seen during a heroin or cocaine "rush." This may explain why heroin addicts have a suppressed sex drive—because the drug already heavily stimulates this region.

There was also significant deactivation in the amygdala for women (less for men), leading the scientists to hypothesize that sex may distract us from events in the environment, even those that might justly invoke fear—perhaps, says Holstege, so that we can have sex "without being bothered by outside stimuli."

All of this activity may have some long-term health benefits. From the University of Bristol comes news that in men who reported the highest frequency of orgasm, the risk of fatal coronary events was halved—possibly because sexual activity offers a cardiovascular workout, or perhaps because men with a robust sex life are just happier and less stressed. Another study shows that college students who have sexual intercourse once or twice a week have 30 percent higher levels of immunoglobulin antibody than abstainers. And another, highly controversial study suggests that sex may have long-term positive effects on mood in women. In a sample of sexually active college females, researchers found that women who had intercourse without condoms had fewer symptoms of depression than those who used condoms or refrained from sex altogether. The scientists are quick to say that they do not recommend eliminating the use of condoms for psychological reasons, as sexually transmitted diseases or an undesired pregnancy would easily outweigh semen's benefits. But the study does suggest that some of the compounds present in semen that can be absorbed through the vagina, including testosterone, estrogen, and prostaglandins, may have an antidepressant effect.

A rush of pleasure; a reduction in stress, depression, and fear. Too bad we so often save sex for the end of the day and not the start.

11

NIGHT AIRS

●

I T'S WELL AFTER 11 P.M., and you should be drifting off. Your partner is sleeping peacefully. But you're wide awake, nursing indigestion from dinner's too-large slab of beef, perhaps, or wheezing from asthma or the congestion of a wicked cold.

A sixteenth-century Italian priest, Sabba da Castiglione, cautioned his followers about the "numerous illnesses that night air is wont to generate in human bodies." You know that your blooming infirmity, whatever it may be, is not due to evil air. But it is true that many ills worsen at night. Fever spikes. Skin irritability flares. Gout, ulcers, and heartburn intensify.

Some of the ailments that stalk us in the hours of darkness are byproducts of the body's nocturnal protective mechanisms. At night many of our daytime defenses slow or shut down — our gag reflex, for instance, and the cilia that sweep clean the respiratory tract. The other mechanisms that fill the protective void, including stepped-up acid secretion and heightened inflammatory reactions, have the potential to wreak their own havoc, aggravating everything from ulcers to psoriasis. Low nocturnal levels of adrenaline and cortisol (which ordinarily help keep breathing passages open during the day) make nighttime asthma attacks hundreds of times more common. Also, late night brings a shift in the workings of the lungs: Bronchia become more hyperreactive, and the bronchial passageways that move air in and out of the lungs shrink in diameter by about 8 percent. For healthy people, this squeeze presents little problem. But for those with asthma, the constriction can reduce air flow to the lungs by 25 to 60 percent, causing the coughing, wheezing, and breathlessness of the disease.

Perhaps this feels more like a simple cold. Two tiny obstructive fists seem to have lodged in your nostrils. At the back of your throat is a sore, scratchy lump that makes it hard to swallow. Earlier today you were a whole, healthy person; now, it seems, you're the wretched host to a germ picked up somewhere, in an elevator or from a child who brought it home from school.

Nighttime is when activity for the body's immune cells is supposed to be peaking. Those glands swelling in your neck are filled with burgeoning populations of white cells known as lymphocytes. They're multiplying all right, but their full flowering against this insidious invader will take as long as a week. At the moment, they're not up to the struggle, and so you sniff and hack your way deep into night.

On average, adults get two to four colds a year; children, about four to eight. Researchers have carefully computed the disruption produced by these illnesses: In a typical year, the 500 million or so cold episodes in the United States cause some 400 million days of missed work and school, and more than 100 million doctor visits, for a total annual cost of up to $40 billion.

What is a cold, anyway? How did your spouse resist the germ that plagues you? Why does a bug register in one person as only a scratchy throat and put another in bed for a week?

The cold earned its name from the link stubbornly forged in folklore between catching a chill and contracting a cold. The Greek philosopher Celsus wrote in the second century A.D. that "winter provokes headache, coughs, and all the affections which attack the throat, and the sides of the chest and lungs."

Modern science deemed the cold-cold link a complete myth—until lately. The conviction that ambient temperature had little to do with susceptibility to infection grew out of a study in the 1950s. Scientists persuaded one group of more than two hundred volunteers to perch in a large freezer for two hours and another, equally large group to sit in their skivvies in a room at 60° F; then the whole crew of four hundred or so was exposed to a cold virus. All of the subjects contracted colds at about the same rate.

A decade later, a similar experiment was conducted after the discovery of the most common cause of the common cold, the rhinovirus (from the Greek *rhinos,* nose). Researchers placed the rhinovirus directly into the noses of inmates from a Texas prison, then exposed their subjects to extreme cold. Neither cold nor warmth, clothing or no clothing, wet hair or dry, affected their rate of infection, prompting the researchers to declare that no further studies were necessary.

Science, however, did not heed the advice, and a new study offers some evidence to support the old folk wisdom. In 2005, a team at the Common Cold Centre in Wales subjected ninety healthy volunteers to a body-chilling ice-water foot bath; ninety others in a control group stayed dry. In less than a week, almost a third of the chilled group had developed cold symptoms, compared with less than a tenth of the control group. The reason? When people are chilled, suggest the researchers, the blood vessels in their noses constrict, shutting off the blood supply to the white cells that fight infection. However, say skeptics, the "chilled" subjects were not tested for the presence of cold viruses; their cold symptoms may have been mainly subjective.

That colds tend to flourish in the chilly months is not because of temperature, but humidity and human behavior, argues Jack Gwaltney, a professor emeritus at the University of Virginia and an expert on the common cold. The rhinovirus's survival depends on moist conditions, above 55 percent humidity. More important, cold, wet days keep children densely packed inside nurseries, schools, and colleges, providing an ideal breeding ground for viruses. People, not weather, are the main problem, says Gwaltney: "The best way to avoid a cold is to live as a hermit; the best way to get one is to see a lot of children." Young noses are the major source of cold viruses. "If your child gets a cold, and you're not immune to that virus," he explains, "there's a 40 percent chance that you, too, will succumb."

Viruses are highly contagious bugs; small doses of only one to thirty particles are sufficient to produce infection. And just a day after being infected, a person can transmit the germ. While the average cold virus is most contagious within the first three days of illness, its particles are shed from nasal secretions for up to three weeks. These particles are surprisingly hardy and enduring. In a paper entitled "Rhinovirus transmission: one if by air, two if by hand," Gwaltney and his colleagues reported that rhinoviruses survive and remain infectious on surfaces—hands, doorknobs, counters—and are most commonly transmitted by finger-to-nose inoculation. During only ten seconds of hand exposure, the virus on a donor's hand is transferred to a recipient's fingers 70 percent of the time. Infection most often occurs when people touch a contaminated object or the fingers of infected individuals, then inoculate their own noses or eyes. The team found that transmission could be interrupted by cleaning the surfaces with disinfectant or even applying iodine to fingers.

Once inside your nose, cold viruses are absorbed by a thin film of mucus that coats the little shelf-like structures in your nasal passages called turbinates. Within ten or fifteen minutes, cilia lining the turbinates move the mucus and its baggage to the back of the throat, where they're swallowed and destroyed in the stomach. In unlucky cases, however, the virus particles carried in the mucus are deposited on your adenoids—lymph glands located above the roof of the mouth and behind the nose, which harbor cells to which the virus may readily attach. Here's the kernel of that thick, scratchy feeling in your throat. At first the adenoidal cells docilely accept the virus; later comes the storm. It takes about eight to twelve hours for a rhinovirus to rev up, complete its reproductive cycle, and produce new viruses. Soon thereafter, the cold symptoms begin.

Ah, the symptoms.

Despite popular belief, that stuffy, blocked feeling in your nostrils does not result from mucus but from the swelling of the turbinates caused by dilating blood vessels. These structures normally swell one at a time on an eight-hour cycle, even when you're cold-free. No one knows why, although some scientists speculate that the cycling may give one nasal chamber a rest while the other carries on the job of air conditioning. During a cold, however, both sides swell at once, and inhalation becomes a stifled affair.

The runny nose of a cold is born of thick mucus produced by "goblet" cells in the lining of the respiratory tract. This mucus is suspended in a watery fluid made from plasma that oozes out of junctions in the cells of the blood vessel walls lining the nose. The plasma carries antibodies and bradykinins—immune system chemicals that stimulate the pain nerve fibers in the nose and throat, causing soreness.

Beware the nose-blow, says Gwaltney: A new finding suggests that forceful nose-blowing may do more damage than good, propelling nasal secretions into the sinuses, where secondary bacterial infection can take hold.

If the tickle of mucus sufficiently irritates the nerve endings in your nasal passages, a message may flash up to the sneeze center in your brain. Thought to be located in the brain stem, the sneeze center coordinates the activity of muscles in the abdomen, chest, diaphragm, vocal cords, and throat, which contract, sending saliva flying out of your mouth and triggering copious nasal flow to wash away the offender.

Coughing can achieve even greater powers of expulsion. In most languages, the word for cough is imitative of a respiratory act so violent it can break blood vessels: *husten, toux, tosse.* First there's the sharp intake of breath, followed by a squeeze of muscles in the diaphragm and abdomen while the vocal cords close the glottis of the larynx for a fraction of a second. Then, as the glottis suddenly reopens, a rapid, powerful flow of air is released from the lungs at speeds of some five hundred miles per hour, blasting out whatever evil need be expelled.

Once considered a simple reflex, coughing is in fact a subtle mechanism. In the respiratory tract, from larynx to lungs, are sensory receptors that are triggered by irritants—mucus, smoke particles, and immune system chemicals. These receptors are the "master switches" of coughing. When stimulated, they fire a signal along the vagus nerve to the cough center, in the medulla area of the brain stem. (Here's where the active ingredients in some prescription cough remedies—opiates such as codeine—exert their calming effect.)

Runny nose, sneezing, cough. As Gwaltney points out, none of these cold symptoms result from any direct damage to the body by the cold virus itself, but from the exuberant action—or overreaction—of the body's own immune response. When Gwaltney and his colleagues biopsied cells of the nasal epithelium during a cold, they found no sign of destruction or injury by the virus. The rhinovirus and others like it produce illness by stimulating the body to do things that are harmful to its own cells and tissues. In fact, says Gwaltney, you can create a full-blown artificial cold in the body without the participation of any virus at all. Here's his recipe:

- A smidgeon of *histamines*—to get the nose dripping, dilate the blood vessels (causing that stuffy feeling), and stimulate the sneeze reflex. (It's not easy to provoke a sneeze, Gwaltney says. He tried tickling the nose of his subjects and giving them a snort of pepper; the only thing that worked reliably was placing histamine right on the nasal epithelium.)
- A pinch of *bradykinins* to stimulate the pain nerve fibers in the throat.

- A dash of *prostaglandins* to kindle a good cough and a head-ache.
- A sprinkle of *interleukin* to cultivate malaise.

Inject all ingredients into nose and wait.

These substances are all natural chemicals that mediate the body's inflammatory response—the first line of defense against injury or infection. Inflammation can occur nearly anywhere, in the skin around a splinter, in the joints (where it's called arthritis), in the brain (encephalitis), or in the lining of the nose (rhinitis). Unfortunately, in the case of a cold, all of this valiant inflammatory activity does not immediately rid the body of a virus. Sneezing and nasal irrigation may help remove dust or pollen from the nose, but not so virus particles; they're safely ensconced inside the actual cells of the nose, which we don't especially want removed. Once activated, inflammatory reactions tend to gain momentum, producing quantities of mucus for a week or so before the infection is resolved.

Your spouse hasn't moved; you lie for a while and watch that free, even breathing with considerable envy. Both of you were exposed to the same party, the same child, the same bed and bathroom cup. Why are you the lucky one?

Take a close look at your life. Perhaps you're more vulnerable because you're suffering from chronic stress, which has been linked with greater susceptibility to rhinovirus-induced colds. Maybe you're short on protein or zinc or vitamin E, which can depress immunity. Or perhaps your overzealous immune response makes you more predisposed to suffer symptoms.

As Gwaltney has discovered, not everyone exposed to a virus—or even infected with it—experiences cold symptoms. When he and his team inoculated with rhinovirus a large group of healthy young adults who previously possessed no antibody to the virus, nearly all of them became infected. Yet only 75 percent went on to develop symptoms of a cold. Just what's going on in the one in four people who don't succumb remains a riddle. "There are some people who claim never to get colds," says Gwaltney. "My wife is one. It's pretty irritating. These people may not be making normal amounts of inflammatory mediators. If this is the case, there's an irony here. We know that cold symptoms are the result of the body's inflammatory response to a foreign invader. So people with more active immune systems may be more prone to developing cold symptoms than people with a less powerful immune response."

Apparently, you're one of the "active" set, so you get up to seek some medicinal assistance. In your bathroom cabinet may reside a crowd of cold remedies, each specifically designed to combat one symptom or another. The decongestants shrink that turbinate nasal tissue to relieve the stuffy feeling. Antihistamines suppress sneezing by acting on the histamine receptors in the sneeze center (where they also produce drowsiness). Ibuprofen helps to ease general malaise. If your shelves are empty of over-the-counter cough remedies — suppressors designed to sit on the tickle, demulcents to soothe it, and expectorants to loosen the phlegm — don't bother making a trip to the pharmacy. A major review of nonprescription cough medicines concluded there is little to recommend them.

Your medicine cabinet certainly holds no real cures for the common cold either; they simply don't exist. The search for one has led into many a stagnant backwater, though few match that of Thomas Jefferson. The great president is said to have counseled a friend to stave off colds as he did, by — of all things — every morning soaking his feet in cold water. A century later, a popular remedy for aborting a cold was to irrigate the nose twice a day with warm water and borax: "No syringe necessary; but by simply immersing the nose in a basin of water, and making forcible inspiratory and expiratory movements, holding the breath at the epiglottis, the nasal passages may be thoroughly irrigated." There are advantages to the syringe from the standpoint of neatness. But the treatment has no proven benefits, Gwaltney says, and may have inherent risks if the borax is contaminated with bacteria.

Modern efforts to cure the common cold have focused on magic bullets targeting a single symptom, but these provide only partial relief. Any really effective cure, argues Gwaltney, must tackle both the virus itself and the symptoms. For more than a decade, Gwaltney has been working on this combination antiviral, anti-inflammatory cold treatment. One such blend of interferon and ibuprofen he tested not long ago seemed promising: In a large group of people with full-blown colds, the team counted and weighed facial tissues from days two to five of their illness and found that the group treated with the combination therapy drug had dramatically reduced severity of sneezing, nasal obstruction, sore throat, cough, headache, and malaise; this group also had 71 percent less nasal fluid and reduced its tissue use by more than half. Treatment must take place early, though, for full benefit.

Gwaltney and others studying colds spend a lot of time measuring mucus. They collect used tissues after nose-blowing, count them and weigh them, then subtract from the total the weight of the same number of dry tissues. "Not particularly pleasant work," says Gwaltney. Nevertheless, measuring soggy tissues has yielded much hard information, including insights into the circadian nature of cold symptoms.

The nose, it turns out, runs by its own clock. Graphs of sneezing, stuffy nose, runny nose, and itchy nose, whether from cold or allergy, show a peak in the morning hours. Volunteers infected with cold and flu viruses use the most tissues in the morning, between 8 and 11 A.M. (with sneezing at its heaviest around 8), and the fewest between 5 and 8 P.M. Cough frequency, too, shows a marked circadian rhythm, peaking between noon and 6 P.M. Most of us would probably fail a quiz on this. I feel certain that I cough more at night. But questionnaires suggest that people are not very good at estimating their own cough frequency; reliable counts can be captured only with tape recorders.

Illnesses of many types are affected by the body's biologic rhythms. As Michael Smolensky, a chronobiologist at the University of Texas, suggests, the daily fluctuations in such ailments as allergies, hypertension, gout, and asthma may be so pronounced that testing for them at the wrong time of day can mean false results. To diagnose an allergy, for example, doctors rely heavily on skin tests. But as Smolensky points out, the skin's response to histamine and to the allergens in house dust and pollen is most pronounced in the evening, just before bedtime, when few doctors schedule appointments. Blood pressure runs higher in the afternoon, so morning diagnostic tests can underestimate the severity of hypertension. A patient may be diagnosed as normal by one physician he sees in the morning and as hypertensive by another who sees him in the afternoon.

If rhythm is as critical a factor in sickness as it is in health, it would seem common sense for doctors to pay close attention to the timing of not only diagnostic tests but also drug treatments. Unfortunately, surveys suggest that many physicians are still unconvinced that circadian rhythms are an important aspect of disease or therapy. This is a problem, say chronobiologists, because the body may process the same dose of a drug in different ways depending on the time of day.

Though direct evidence of circadian modulation of drug action is rare, one groundbreaking study in 2006 showed that in mice, at least, the circadian clock drives rhythms in the genes that allow the body to respond to drugs and other foreign substances. Mice with normal clocks cleared their bodies of the drug pentobarbital much faster at night than during the day. Mice with mutated clocks had severe deficits in clearing the drug from their systems at all times of day. They also experienced more toxic side effects to two chemotherapeutic agents.

Research in humans suggests similar circadian effects. One study shows that anesthesia for dentistry lasts longer in the afternoon than in the morning: Lidocaine taken between 1 P.M. and 3 P.M. relieved dental pain for three times longer than it did early in the morning. On the other hand, a 2006 report showed that patients who undergo anesthesia for surgery in the afternoon suffer more pain and postoperative nausea and vomiting than those who receive the drugs and have their procedures in the morning. This may result from errors in the delivery of the anesthesia because of physician fatigue, but it may also be rooted in the way the body handles a particular drug at a particular time of day.

The pace at which drugs work in the body—the way they're absorbed, metabolized, and excreted—is shaped by circadian rhythms in a variety of body functions. The daily rise and fall of different hormones affect drug absorption. The rhythmic nature of stomach activity (emptying faster by day, more slowly by night) means that some drugs taken by mouth before bed may move into the bloodstream more slowly. Medications taken late in the day are in general more rapidly destroyed by the body because higher body temperature speeds the chemical reactions the body uses to detoxify a foreign substance. Such time-of-day effects have been documented for more than a hundred drugs.

The goal in timing the delivery of medication, says the chronobiologist Russell Foster, should be to balance what the body does to the drug with what the drug does to the body. This is especially true with cancer drugs, where timing can mean the difference between life and death.

• • •

About twenty-five years ago, my mother underwent courses of radiation and chemotherapy for cervical cancer. The case was an extremely aggressive one: She was diagnosed in February and died in July. In the early days of her treatment, she suffered nausea and loss of appetite. I tried to entice her to eat by offering her small portions of her favorite foods, a perfectly seared bratwurst or cheese from the new gourmet shop down the street, even brownies laced with hashish, which has been known to relieve nausea. Nothing worked. My mother's intestinal distress resulted from the lethal effect of the highly toxic chemotherapy drugs on the swiftly dividing cells lining her gastrointestinal tract.

The aim in cancer treatment is to kill tumor cells without killing normal cells. Many anticancer drugs are designed to destroy only cells that divide rapidly. Since cancer cells multiply at a faster pace than most normal cells (every six to twelve hours, versus every twenty-four hours), they're preferentially destroyed. But chemotherapeutic drugs are imprecise weapons. They strike not just their intended target but also innocent bystanders, the normal, healthy cells in the body that also happen to divide quickly, such as the cells in the bone marrow, hair follicles, and in the lining of the digestive tract. This is what causes the side effects of the therapy—anemia, hair loss, and gastrointestinal upset. The toxicity of these drugs limits how much and how often they can be taken.

Francis Lévi believes that *when* people receive anticancer treatment is at least as important as what they receive in determining whether the treatment will be successful or dangerously toxic. Lévi, a physician who studies circadian interactions with cancer at the Hôpital Paul Brousse near Paris, belongs to a growing group of researchers who believe that timing is critical to successful cancer treatment.

Most cancer patients still receive chemotherapy at times that are convenient for the hospital staff. But more and more studies by Lévi and others are showing that giving cancer drugs at carefully selected times of day can maximize their therapeutic effects and minimize their toxic side effects.

The key is understanding the distinct timing of cell division in cancer cells and normal cells. Take lymphoma. Cells in certain kinds of lymphoma tend to divide between 9 and 10 P.M., the cells in the gut lining at around 7 A.M., and those in the bone marrow near noon. William Hrushesky, a pioneer researcher in chronotherapy at the University of South Carolina School of Medicine, has found that the cells lining the gut proliferate twenty-three times as much during day as they do at night. So a chemotherapeutic agent that's known to damage the gut and bone marrow might be expected to be less toxic—and more effective against lymphoma cells—if given in the nighttime hours.

More than two decades ago, Hrushesky published a study on the timing of chemotherapy in forty-one women with ovarian cancer. Women on one schedule developed only half the negative side effects as women on another schedule. According to Hrushesky, every measure of toxicity was lowered several-fold depending on what time of day the drugs were given. "Those women who received the drugs at least damaging times of day also had a fourfold better chance of surviving five years," says Hrushesky. "This shows that human cancer susceptibility to chemotherapy depends on the time of day these drugs are received."

More recently, Francis Lévi has had similar success treating advanced colorectal cancer with a drug called oxaliplatin. In one study, Lévi found that when the drug was delivered in a conventional, steady dose, tumors were reduced by 30 percent; with chronotherapy, by 51 percent. "And," adds Lévi, "the most effective therapy was also the least toxic one," with less severe side effects.

How common is the practice among oncologists of carefully timing the delivery of cancer drugs? "Ten or fifteen years ago, most people thought we came from another planet," says Lévi. "They're listening now, but it's a slow process of acceptance."

Fretting about whether the well-timed delivery of cancer drugs might have saved my mother—or at least eased her discomfort—is the sort of speculation that keeps me up at night, cold or no cold.

It's midnight, day's official close. If you're like me, you would give anything to slip into the gray shadowland of sleep. We delight in thresholds, but not if the crossing is protracted. Think of the painful lag Brick feels in *Cat on a Hot Tin Roof* as he waits for that booze-induced "click" that ushers him into welcome oblivion; think of the intolerable waiting of adolescence, as Theodore Roethke calls it, "a longing for another place and time, another condition"; think of this, the long, fitful effort to fall asleep. Normally I drop off easily as soon as my husband switches off the light, as if there's a cable linking his lamp with my brain. But some nights, circumstances conspire against it: indigestion, a cold, or just a mind buzzing in circles like some heat-maddened fly.

Anxiety and stress are, in fact, among the chief reasons people have trouble getting to sleep. There are medications, of course, including a whole new generation of "gentle" somnolents. But experts say that sleeping pills of any kind derange normal sleep. The best strategy is good "sleep hygiene": going to bed at a regular time, covering your clock, not exercising late at night. It also may help to limit demanding mental work and, perhaps, to heed the advice of Robert Burton, "Hear sweet Musick . . . or read some pleasant Author" and hope for a quick crossing over.

12

SLEEP

●

T HE ITALIANS CALL IT *dormiveglia;* the Germans, *einschlafen.*
But the English language possesses no single, really expressive
word for the passage into sleep. I don't know why this is so. Perhaps for the same reason that we have a perfectly good word for birthday but none for life's other bookend. Perhaps it reflects the significance our culture ascribes to its perceived moments of importance.

"Sleep is the most moronic fraternity in the world," wrote Vladimir Nabokov, "with the heaviest dues and the crudest rituals." Because
we lose consciousness so dramatically when we doze, it was assumed
for centuries that sleep shut off the brain. Slumber was a kind of passive suspension of mental activity, a dark slice out of time—an idea
that held on well into the twentieth century.

Now we know that sleep is a remarkable journey of five stages repeated cyclically over the course of the night. These are richly varied states involving shifts in brain waves, body temperature, and biochemistry, in muscle and sensory activity, in thoughts and level of
awareness. Though the depth and quality of the journey may differ
from person to person, from age to age, the pattern is more or less the
same: four or five such cycles, each about an hour and a half long, alternating between quiet deep sleep and active REM sleep.

The brain is hardly "absent" at any time during the passage, says
Jerry Siegel, a sleep researcher at UCLA. "It actively puts itself to sleep,
then self-activates during sleep. As many neurons switch on in sleep

as switch off; what changes is their overall pattern of activity." While some brain areas may quiet down compared with their daytime activity, others fire up. Even during deep sleep, "when consciousness may be totally obliterated," adds J. Allan Hobson of Harvard Medical School, "the brain is still roughly 80 percent activated and thus capable of robust and elaborate information processing."

Even after the discovery of sleep's cyclical nature, the belief prevailed that sleep was mainly downtime; its purpose, to cure sleepiness. Only now are we waking up to its astonishing complexity and the range of ways it affects body and mind. In maintaining good health, says William Dement, sleep may be more critical than diet, exercise, even heredity.

Sometimes when I can't drop off, I climb out of bed and wander into my daughter's bedroom. I'm a notoriously poor observer of my own slumber, so it's enlightening to watch hers. One hand cradles her face. Her breathing is light but regular. Though it's warm in the room, she's snuggled under the covers because her body temperature is on the wane.

Did sleep come upon her like that sly fox I saw, stealing forward, pausing, stealing back? Until recently, science described falling asleep as a slow diminishment of alertness over several minutes as the brain passed from 100 percent wakefulness to 100 percent sleep. Now it appears that drifting off is not a gradual process at all but a sudden leap, a quick neural shift from consciousness of the outer world to nearly complete sensory blindness.

The shift, it turns out, is executed by a sleep switch in the hypothalamus of the brain. Baron Constantin von Economo, an Austrian neurologist, first identified the area of the brain where the switch exists in patients with encephalitis lethargica, a form of sleeping sickness that swept through Europe and North America in the 1920s. Most of Economo's patients slept excessively, twenty hours or more a day. The sleepy patients, he discovered, had lesions in the hypothalamus.

Scientists have recently pinpointed the switch, a cluster of neurons that work together to turn off the brain's arousal circuitry. It's what an electrical engineer would call a flip-flop switch, explains Clifford Saper, a neurologist at Harvard Medical School. Such switches are designed to produce "discrete states with sharp transitions and to avoid transitional states," says Saper. "This flip-flop circuit model might explain why wake-sleep transitions are often relatively abrupt (one 'falls' asleep and suddenly wakens)." People with narcolepsy behave as if their switches are destabilized, easily dozing off during the day and waking more often at night.

The cells in this switch cluster are sensitive to environmental factors such as heat, which may explain why a warm bath or a hot day causes drowsiness. But scientists suspect that the switch is primarily driven by those same two interacting mechanisms at play during the afternoon doldrums: the homeostatic pressure for sleep and the circadian alerting system. In the evening, the latter sends such a powerful alertness signal that it creates a "wake maintenance zone" between 6 and 9 P.M., when it's hard to get to sleep even after severe sleep deprivation—except for extreme larks. People sleep best if they go to bed two or three hours after this zone, when the pressure for sleep is heavy indeed. By this time, the circadian alerting system has started its slide toward night, and the body's master clock has signaled the pineal gland to boost its production of melatonin, telling us it's dark, time for slumber.

When sleep seized my daughter, her skeletal muscles relaxed, and the favorite stuffed panda she clutched dropped from her hand. If she had been wearing the Medusa's head of electrodes that sleep labs use to record electrical activity in the brain, the delicate, nervous marks of the pen inscribing moving paper might have recorded a shift from the alpha waves of drowsiness, like the regular teeth of a comb, to the lower-frequency theta waves of half-sleep or early sleep.

Sometimes at this early stage, the peculiar floating or falling feeling one may experience while going to sleep is interrupted by a brief spasm known as a hypnic jerk or myoclonic twitch—a quick muscular contraction in arms, legs, sometimes the whole body—which startles one back to wakefulness. This is more frequent in adults than in children, and more common in people who are nervous or overtired. Some evolutionary biologists speculate that the hypnic jerk may be a reflex left over from our arboreal ancestors—useful in avoiding a slip from a sleeping perch.

Even in the half-sleep of stage 1, my daughter would not hear my murmured "good night" or smell supper's lingering aroma of roasted potatoes. Her ability to take in signals from the outer world has slipped to nil. Her mind may wander from the focused thoughts of wakefulness—the words she learned for her vocabulary test, the impending visit by her grandmother—to more associative thinking and then to actual pictures that change rapidly, so-called hypnagogic hallucinations, which leap from subject to subject and from one outlook to another. If I nudge her awake to ask if she's asleep, she would likely deny it.

Not so after a few more minutes. As her sleep deepens, a polygraph would at first bristle with the fleeting sleep spindles (short bursts of EEG activity) and so-called K-complexes of stage 2 sleep, then shift into the longer delta waves of stage 3, and finally into the synchronized oceanic waves of deep, slow-wave, stage 4 sleep. The neurons in her brain that do their own individual thing when she's awake would begin to fire synchronously, producing those big, slow waves.

At this stage, I would have to shake her hard to wake her. Her breathing is slow and regular; her muscles are limp. Her pituitary gland may have begun releasing surges of essential hormones, including gonadotropic hormones, which play a role in the development of her sex organs, and growth hormones, which spur her cells to divide and multiply. Most of her bone growth may be taking place now. When scientists implanted tiny sensors into the shinbones of lambs and measured bone length every three minutes for three weeks, they found that 90 percent of the growth occurred when the lambs were sleeping or lying down at rest.

This stage of deep sleep may last a half hour to forty-five minutes before the spindles and K-complexes of lighter sleep again reappear. How long the stages of deep sleep last for any individual may be affected by genes. When Swiss researchers studied a gene that regulates adenosine in a group of more than one hundred student volunteers, they found that the 10 percent of students who had a mutation in the gene gleaned an extra half hour of deep sleep and reported waking up less often than those students without the mutation.

Moving through the phases of sleep is a bit like diving, plunging into deeper and deeper waters. If I sit and watch my little diver long enough, I can see her suddenly resurface. She shifts position from back to stomach or turns from side to side. Her breathing and heartbeat quicken as if she were in the throes of movement or emotion. Cells fire in the motor region of her brain, but a complex system of neurotransmitters stops these brain signals from getting through to the motoneurons that actually activate her skeletal muscles; instead, these muscles become so thoroughly relaxed they're virtually paralyzed. Only the muscles of her eyes are unaffected, and her eyeballs flit about wildly beneath her lids. Perhaps most astonishing, her brain's control over essential physiological processes diminishes, including the maintenance of her body temperature and proper levels of blood gases. A polygraph would reveal jagged theta waves, punctuated with short bursts of alpha and beta waves, as her neurons begin to fire as busily—and as individually—as they do during waking.

This is REM sleep, a bizarre state that in some ways resembles wakefulness more than sleep. For the next five or ten minutes, my daughter's body closes down, but her mind is off on a singular adventure of its own, seeing and hearing things that aren't there. About a quarter of a night's slumber is occupied by REM sleep, in four or five bouts a night, building in duration from episodes of about ten minutes at the start of a night to thirty minutes as dawn approaches. This is the time of intense dreaming. If I shook my daughter's shoulder now and asked her what was going through her mind, she might describe a dream of feckless flying or sailing down a stream of ink.

Everyone dreams, says J. Allan Hobson; any impression to the contrary is likely rooted in poor recall. We dream in both REM and non-REM sleep, but non-REM dreams tend to be short, fragmented, and dull. REM hosts the kind of delirious reverie characterized by strange, vivid hallucinations, illogical thinking, emotion, and confabulation.

Recent advances in imaging have given us a remarkably clear picture of where dreaming activity takes place—which parts of the brain are revved up and which are quiet. Silent are the regions of the prefrontal cortex known to be important in working memory, attention, and volition. The neurotransmitter systems required for these functions during waking—specifically, serotonin, histamine, and noradrenaline—are simply shut off during REM sleep, says Jerry Siegel; with them go insight, reasoning, and a logical sense of time. Lively are the cortical regions essential to visuospatial processing, including the hippocampus, the center for cells devoted to a sense of place and direction. This may contribute to the "virtual navigation" apparent in dreaming. Also active are the amygdala and the limbic system, both central to the feelings of anger, anxiety, elation, and fear that so often accompany a dream story.

Unlike my daughter, whose dreams—at least the ones she remembers—are nearly all sweet, I had nightmares as a child, ominous and terrifying. Tore Nielsen, of the Dream and Nightmare Laboratory at the Hôpital du Sacré-Coeur in Montreal, suspects that nightmares may in some cases arise from circadian rhythm disturbances—from

REM activity that is phase-advanced (occurring earlier in the cycle than normal), a possibility that deserves study, he says.

In his investigation of dreams, Nielsen has found that women typically have more nightmares than men. In a group of more than a thousand university students, women reported two nightmares a month (men, about one and a half) and a higher prevalence of dreams with frightening themes. "This gender difference is quite robust," says Nielsen, "first appearing early in adolescence and measurable right into old age. The disturbing dream content may be a function of female biology—for example, monthly hormonal fluctuations— or it may be due to sociocultural influences that differentially affect women, such as traumatic experiences, depression, and sleep disorders."

The nightmare I remember best from childhood was set in a pale pink house around the corner from my home. The man and woman who lived there hovered on a balcony above the sidewalk where my mother stood, calling her name. When my mother looked up, the couple poured laundry detergent in her eyes, a steady stream of blinding white powder. I tried to scream, but the sound never left my body. And when I tried to come to her aid, I found that all my muscles were paralyzed (as they were, in REM sleep). I awoke with a start and lay there immobilized, heart pounding violently. It was just a dream, I told myself. But I could not flush from my mind the image of my helpless mother blinded by Cheer, and had to assure myself by padding down to her bedroom to see her peacefully asleep.

This dream was vivid and vividly remembered. Why are so many others extinguished? How many mornings I've awoken with the remnants of a dream buried deep or scattered like bones at an ancient ruin. Its residue remains, but the dream story is gone. "REM dreams tend to be forgotten if they are followed by extended periods of non-REM sleep," says Allan Rechtschaffen, the retired head of the University of Chicago Sleep Research Lab. The dreams we remember best are those arising in the final REM period from which we awaken.

On average, we dream for an hour and a half to two hours each night, with four or five distinct dreams. If we live out our allotted seventy-five years or so, that means we'll spend about six years vividly dreaming, for a lifetime total of some 100,000 to 200,000 dreams.

Right now I would take one brief reverie. Back in bed, the clock flicks to 12:38 A.M. Exhaustion is doing battle with tension; my physical self, fatigued from the day's activity, is craving slumber, but my mind is alert like some wary crane. Exercise doesn't always lead to a good night's sleep, but it does tend to improve sleep patterns for insomniacs (at least as well as sleeping pills, according to some studies), and even in good sleepers it can modestly increase sleep length and depth. Some scientists suspect that the positive impact of exercise on sleep may be enhanced by the light exposure that often comes with it. A daily dose of strong natural light has been shown to have both sleep-promoting and antidepressant effects. Sedentary adults typically get about twenty minutes of daily exposure to natural light, while people who exercise get about three times that amount.

I might turn to a nightcap for help drifting off, but the practice is not recommended. Though a drink before bed may initially make one feel drowsy, once the alcohol has been processed by the body, the resulting compounds have a stimulating effect, disrupting slumber later in the night. Researchers recently glimpsed why this might be so. Alcohol affects the thalamus, a region of the brain integral to sleep-wake rhythms and to the spindle waves that occur during stage 2 sleep. So sensitive is the thalamus, researchers say, that just a drink or two will make for lighter sleep in the middle of the night, or even full wakefulness.

I know I'll eventually drop off. I'm not like the incurable insomniac for whom night is no downward slope to certain oblivion, but rather a flat and despairing palm of wakefulness. Still, I have too many demands the next day to lose even an hour's rest.

Losing sleep didn't always feel so—well—so *punishing*. Once, when I was asleep in a lean-to hut in the mountains of New Hampshire, some night animal rustled or hooted me awake. I slithered to the edge of the wooden platform and lay there in my sleeping bag, watching a crescent moon slowly slide across the sky and then set over a pine-edged slope. The night was all stars and dew and strange perfumes, and the hours passed by lightly, with faint shifts of temperature and luminosity. I felt energetic, awed, happy to stay awake through the night to witness its transit.

But I was young then. Now I'm a working mother jealous of my slumber. So what do I do? Fret, brood, lose sleep. And what of it?

Sleep is sore labor's bath, nature's soft nurse, the balm of hurt minds. It is Nabokov's moronic fraternity or Coleridge's gentle thing, belov'd from pole to pole. The number of metaphorical descriptions of what sleep *is* is nearly matched by the number of theories about what it *does*. In the past few decades we have made good progress teasing apart the physiology and neural architecture of sleep, but the question of its purpose remains a biological dilemma of the first order. What could possibly be the advantage of so complicated and dangerous an undertaking as shutting down our sensory systems, paralyzing our muscles, putting ourselves at risk for a third of our hours? Wouldn't it be far better to be ever ready, up and running?

"If sleep does not serve an absolutely vital function," Allan Rechtschaffen once noted, "then it is the biggest mistake the evolutionary process has ever made."

This vital function, however, has proved devilishly elusive. One common technique to reveal the job of a bodily organ or behavior is to take it away. Rechtschaffen and his colleagues at the University of Chicago conducted a famous series of experiments showing that rats deprived of sleep eat more than usual but still lose weight and double their energy expenditure. They lose control of their body temperature and develop sores on their paws and tails that don't heal. After about two and a half weeks, they die—faster than if totally deprived of food.

For obvious reasons, no such experiments have been conducted on humans. But in 1965, a high school senior named Randy Gardner deprived himself of sleep for 264 hours to break the world record as part of a project for the San Diego Science Fair. After the allotted eleven days of continuous wakefulness, Gardner suffered no psychosis and no serious medical problems, but he did show deficits in concentration, motivation, and perception—as did William Dement, who observed him. Dement began spending nights at Gardner's house on the second day of the experiment to make sure the young man stayed awake and to monitor his physical and mental health. On the fifth day of his own wakefulness, Dement drove his car the wrong way up a one-way street and almost collided with a police car.

Such cases of sleep deprivation are extreme, of course. "There's another long-term, massive experiment in minor sleep debt going on in contemporary culture," says Charles Czeisler. A 2005 poll taken by the National Sleep Foundation found that about 40 percent of Americans get less than seven hours of sleep a night during the work week. That's an hour or two less than people got fifty years ago. Moreover, one in six people reported sleeping less than six hours a night—a substantial curtailment of sleep that may have serious consequences.

Consider the case of my friend Harri, a vibrant woman in her early fifties who teaches reading. One spring morning, Harri woke up with profound amnesia. She couldn't recall anything she had done or said the previous evening. She remembered making dinner and sitting down with her family to eat, but the span of hours after dinner was a blank.

Harri doesn't drink or smoke or take any drugs and considers herself in good health. The memory lapse spooked her a little, but she wrote it off to stress. Then it happened again; then twice more in one week. One evening, when her teenage son called from college, she spoke with him at length, but the following morning she had no recollection of the conversation. Her son told her later that she had been so unresponsive on the phone that he had yelled, *"Get Dad!"*

Finally, Harri consulted a neurologist. "I just wanted to know that it wasn't a brain tumor," she said. "I thought I could deal with just about anything else." At first the doctor suspected epilepsy, but a battery of EEGs and other tests proved normal. Then she inquired about Harri's sleep habits. Ever since Harri could remember, she had slept only five hours a night. "It's not that I wasn't tired," she said. "I just felt I should squeeze as much as I could into the day by stealing an extra two or three hours to work."

I can understand the temptation. By staying up just one more hour each day for, say, seventy years, one could add to life 25,550 hours of reading time. But cheating sleep comes at a price. The result of Harri's cumulative sleep deprivation, the doctor told her, was a pathological sleepwalking state. After dinner, her brain dozed while her body continued to go through the motions of the evening.

Following the prescription was simple: Get at least seven hours of sleep. This Harri does now, religiously, and her amnesic episodes have disappeared.

"Overmuch" sleep "dulls the spirits," wrote Robert Burton in his *Anatomy of Melancholy*, "fills the head full of gross humours; causeth distillations, rheums, great store of excrements in the brain, and all other parts." As recently as a decade ago, eminent researchers in the sleep field were actively engaged in convincing people that they need not sleep long to sleep well. The idea was that only a "core" of four or five hours of sleep was necessary; the remaining three or four hours we spent in bed was optional, a luxury or hedge against future overwhelming physiological demands, but not required for optimal brain function. The extra hours of sleep were like the excess pounds of body fat we carry as anachronistic protection against periodic bouts of famine; in our current mode of living, we just don't need them anymore.

Of late, scientists have unequivocally overturned this belief. Though just how much sleep is best is still debated, evidence is mounting to suggest that for most of us, between seven and eight hours is optimal; less than six is simply not enough. David Dinges and his colleagues at the University of Pennsylvania found that people who slept less than six hours a night for two weeks took a dive in cognitive performance equal to two nights of total sleep deprivation. Though the sleep-restricted subjects reported feeling only slightly sleepy, they did poorly on nearly all tests of alertness, attention, coordination, and cognitive tasks. Those who slept only four hours a night for the same two-week period had total lapses like my friend Harri's, when they simply failed to respond to a stimulus.

Other studies have reported that in terms of sedative effects and performance impairment, losing two hours of sleep out of a regular eight is tantamount to drinking two or three beers; losing four hours is like drinking five beers; and losing a whole night's sleep is like downing ten.

Sometimes we can compensate for the deprivation. Feelings of sleepiness may be offset by the alerting effects of our circadian clock and the stimulus of excitement, interest, or stress. And some studies suggest that sleep loss can actually increase activity in parts of the cortex not normally involved in a given task, suggesting that the brain may call in reinforcements to counter the dulling effects of sleep deprivation.

But there are limits. "With one night of short sleep, some people don't feel much impact," says Czeisler. "But everyone suffers after a week or two because the effects are cumulative. One week of restricted sleep builds a level of impairment equivalent to 24 hours of consecutive wakefulness. Two weeks is like 48 hours."

With prolonged sleeplessness, people often fall quickly into micro-sleeps, those sleep episodes of three to ten seconds that push through wakefulness. When you're driving down the highway at sixty miles per hour, that's enough time to veer off track hundreds of feet—say, across the median strip. The National Highway Traffic Safety Administration (NHTSA) estimates that drowsiness increases a driver's risk of a crash or near crash by at least a factor of four and that drowsy driving is responsible each year for at least 100,000 accidents and 1,500 fatalities.

Drowsiness is considered a major cause of accidents in virtually all forms of transport, surpassing alcohol and drugs. As William Dement points out, most people don't know that sleep deprivation, not alcohol, was a key factor in causing the *Exxon Valdez* to run aground in 1989, spilling millions of gallons of crude oil. The shipmate in control of the tanker at the time was operating on only six hours' sleep over two full days—not enough to stay alert, according to the National Transportation Safety Board. This was also the case in the tragedy of 2002 in Webbers Falls, Oklahoma, when a barge struck an interstate highway bridge, killing fourteen people, and in the explosion of the space shuttle *Challenger*. The night before the *Challenger* launch, NASA managers had gotten less than two hours of sleep. The report on the disaster suggested that sleep deprivation may have impaired their judgment about launching the rocket despite temperatures too cold to allow the O-rings to function properly.

How could the experts have been so wrong in the past about the devastating cognitive effects of accumulated sleep debt? In part, because scientists relied on subjects' self-reported feelings of sleepiness rather than on objective performance tests, says Czeisler. We are poor judges of our own sleepiness and its impact on our functioning. We tend not to recognize the signals of serious fatigue or realize the effect it has on how we listen, read, calculate, talk, operate machines, or drive. "And most of us," adds Czeisler, "have forgotten what it really feels like to be awake."

The mischief done by shortchanging sleep goes well beyond the workings of the mind. Eve Van Cauter, a sleep researcher at the University of Chicago, has found that restricting sleep to four or fewer hours for successive nights results in widespread changes in the body, including some that cause illness and mimic the hallmarks of aging.

To parse the popular belief that losing sleep increases one's chance of getting sick, Van Cauter examined the effect of sleep restriction on the body's immune response to vaccination. She and her team administered flu vaccines to a group of twenty-five volunteers who had slept only four hours a night for six nights. Ten days after the vaccination, their antibody response was less than half that of normal sleepers.

Van Cauter also found that sleep loss impairs the body's ability to perform basic metabolic tasks, such as regulating blood sugar and hormones, creating changes that closely resemble those of aging. After several nights of sleep restriction, a group of eleven lean and healthy young men showed signs of trouble processing blood sugar, causing a condition that looked like early diabetes. The ability of their insulin to respond to glucose was reduced by a third, and they took 40 percent longer than normal to regulate their blood sugar after a high-carbohydrate meal. The subjects also showed high levels of cortisol during the evening hours, when the hormone is supposed to ebb. This late-day spike in cortisol—a risk factor for hypertension—is typical of much older people. In some subjects, the severe sleep restriction had the effect of making the body of an eighteen-year-old look like that of a much older man.

Most of us know from experience that when we're tired from too little sleep, we tend to eat more. Van Cauter and her team have discovered an explanation: Sleep restriction reduces the body's supply of leptin, the hormone that signals satiety and regulates energy balance. She and her team have reported that subjects whose sleep was restricted to four hours a night had 18 percent less leptin in their blood and 28 percent more of the "hunger hormone" ghrelin than those sleeping seven or eight hours a night. They also felt hungrier and had more appetite for calorie-dense carbohydrates such as cake and bread. The body seems to respond to a loss of a few hours' sleep in the same way it responds to a deficit of about one thousand calories—by cueing its systems to slow metabolism, deposit more fat, and step up appetite, particularly for high-calorie foods.

In fact, Van Cauter suspects that the epidemic of insufficient sleep in our society may be responsible for the epidemic of obesity. Indeed, in 2005 researchers showed that obesity is tightly correlated with sleep time. The survey of 9,500 people from across the United States, aged thirty-two to forty-nine, revealed that those who reported sleeping five hours a night were 60 percent more likely to be obese than their counterparts who slept seven hours or more.

A twist to this tale comes from a 2006 report on the sleep habits and weight gain of 68,000 women over a sixteen-year period. Like the women in the earlier study, the subjects who slept five hours or less were more likely to gain weight over time than those who slept seven hours—but not because they ate more or exercised less. In fact, on average, the short-sleepers in the study consumed fewer calories and had roughly the same level of physical activity. The culprit, instead, could be a lower metabolic rate or reduced nonexercise activity thermogenesis—the NEAT movements we saw earlier, such as fidgeting—that may come with sleep restriction.

If you doze, you lose time for work and play. If you don't doze, you lose the ability to concentrate, react quickly, and fight off infection. You also put yourself at greater risk for diabetes and obesity, high blood pressure and heart trouble.

It's enough to keep anyone awake with worry.

However, as Allan Rechtschaffen says, "The effects of sleep deprivation alone will not tell us the function of sleep." Saying that sleep is necessary to stay alert, awake, and healthy is inadequate, says Rechtschaffen. "We have not even begun to understand what it is about the physiology of sleep that is necessary to prevent the effects of sleep deprivation. Would we be happy with the conclusion that the function of eating was to prevent an increase in appetite? We need to know much more."

Some hints have come from surprising sources: the giraffe and the shrew, for instance. Jerry Siegel and others have delved into the sleep of dozens of species to probe its purpose. All animals sleep, says Siegel. There's the watchful one-eyed sleep of certain birds, which lets them scan their surroundings for possible signs of danger while they snooze. Dolphins sleep half a brain at a time, allowing them to swim and control their breathing voluntarily while getting restful slumber. First one hemisphere sleeps for a couple of hours, then the other, until the dolphin has satisfied its sleep needs.

In scrutinizing animal sleep, Siegel has found some unexpected patterns: "The amount of time an animal sleeps and the time it spends in REM sleep vary enormously over different animal species," he says, "even among species of the same order." This is surprising, because closely related species usually have a similar brain structure and DNA, and so would be expected to have similar sleep habits. There are some general rules, however: Herbivores, who must eat throughout the day to fill their needs, have shorter sleeps; carnivores, who can get a meal in one fell swoop, have longer ones. Omnivorous humans fall in the middle. For Siegel, the way nature has adapted sleep to the living conditions of an organism suggests that sleep's core function may be to help a creature exploit its specific ecological niche.

Whatever functions occur during sleep, says Siegel, may have migrated to the quiescent period of a twenty-four-hour day because it's efficient to perform them then. One such task is the repair of metabolic damage done to the body in wakefulness. "Sleep duration is linked to body size," Siegel explains. "The bigger the animal, the shorter the sleep." Giraffes and elephants sleep two to four hours a day; armadillos and opossums about eighteen hours. Siegel suspects this has something to do with the high metabolic rate of small animals, which generates more cell damage, requiring more of sleep's restorative effects.

Repair during sleep may be especially important for the brain. Sleep may give the brain an opportunity to mend itself and to do "housekeeping" chores, such as restocking proteins and strengthening synapses, says Siegel. To achieve these tasks, the brain must shut down so that the metabolic activity of neurons doesn't get in the way. The lower brain temperature and slowed metabolic rate that accompanies deep sleep may allow enzymes to more efficiently repair and rejuvenate cells.

Most scientists believe that sleep has more than one function and that REM and non-REM sleep each plays a different, important role. Humans and animals made to go without one kind of sleep or the other will subsequently make up the debt in that particular type of sleep. The more sleep-deprived people are, the more quickly they will sink into deep non-REM sleep. If deprived of REM sleep only, they will almost immediately pop into REM that is more intense than normal REM, with more frequent eye movements.

Siegel suspects that one key to REM's work may lie in the systems of brain cells that are quieted during this sleep state. Two neurotransmitter systems that are halted during REM, noradrenaline and serotonin, are the ones normally active in waking, enabling body movement and heightening sensory awareness. Siegel theorizes that the shutdown of these systems maintains their sensitivity and smooth functioning during daytime hours. This may explain the uplifting effect of REM deprivation on mood in people with depression. Depriving the brain of REM sleep—that is, allowing the serotonin and noradrenaline systems to keep producing their chemicals—boosts the amount of serotonin available to cells. This is the same principle at work in antidepressant drugs such as Prozac and Zoloft.

Another school of thought suggests a more revolutionary role for REM: quite literally, to make up the mind.

The first time I observed REM sleep was when my infant daughter dropped off after nursing, her head curled into my chest, her hands spread starfish-wise. Almost immediately, I could see the little flickering eye movements of REM. Babies usually spend four times longer in REM than do adults, about eight hours every day. "Indeed, most animals born immature sleep a lot and have a higher total REM time at birth and throughout life," says Siegel.

Why? One theory holds that REM helps establish brain connections during crucial periods of development. The idea is this: At birth, the brain has far more neurons than it needs. As it matures in infancy, it prunes redundant cells and connections in the cortex to strengthen key networks. Cells that are inactive are eliminated. Babies' brains, like those of adults, need periods of deep, quiet sleep for recovery and restoration. But such sleep inactivates brain cells. So, the theory goes, REM steps in to keep neurons in crucial networks active in the resting baby's mind to rescue them from the shearing.

But then why does REM sleep continue through adulthood?

Perchance to learn.

When I was in high school, there was a fad for studying for exams overnight by listening to taped recitations of facts and figures. These were supposed to soak magically into our slumbering brains in time for morning retrieval. So-called sleep learning—acquiring new knowledge while we're asleep—has been a focus of experimentation for decades. In the 1940s, a scientist reported having eradicated the nail-biting habits of almost a third of the boys at a summer camp by admonishing them in their sleep thousands of times over the course of eight weeks. More recently, a group of Finnish researchers claimed to have taught full-term human newborns to discriminate between similar vowel sounds while they were fast asleep, which suggested to the scientists that "this route to learning may be more efficient in neonates than it is generally thought in adults."

But most scientists agree that learning during sleep—that is, actively acquiring new knowledge—is probably impossible. Certainly, attempts to teach slumbering adult subjects vocabulary or foreign languages or lists of items have failed miserably. But mounting evidence supports the idea that sleep—either REM or slow-wave or both—may be essential to effective learning and the formation of memories after the fact, that the sleeping brain processes, sorts, and stores information it gleaned during waking hours.

"Most of us know we need to get a good night's sleep in order to be at our best in learning a task or taking a test," says Charles Czeisler. "What many don't realize is that the sleep we obtain the night *after* learning the task is critical for learning it well, for consolidating the memory of that task."

Robert Stickgold and his colleagues at Harvard recently demonstrated that a night's sleep after learning either a visual task or a motor skill was crucial to improved performance. Subjects taught the visual task could not better their performance beyond a certain level without "sleeping on it." Those taught the motor task showed a 20 percent increase in speed after a night of sleep, while an equivalent period of wakeful time offered no benefit.

Some years ago, scientists found that the same parts of the brain activated while people learned a task were again activated during REM sleep. It's called the replay phenomenon. One recent study showed that the precise part of the hippocampus that fired up during learning a spatial task—finding one's way through a virtual town—was again active during slow-wave sleep.

Perhaps sleep allows the brain to review neural connections made during the day. Or perhaps it serves to reset their strength, restoring the brain's homeostasis, posits neurobiologist Giulio Tononi. In 2004, Tononi and his colleagues reported finding that the specific parts of the brain used during a daytime learning task showed heightened slow-wave activity at night. The team asked volunteers to perform a task requiring complicated hand-eye coordination before they went to sleep. The task is known to demand activity in a particular region of the right parietal cortex of the brain. After training, the subjects slept while their brain activity was monitored with MRI imaging and EEG readings from 256 electrodes on their scalp. The results showed that during post-training sleep, slow-wave activity increased only in that region thought to be activated by the task. Moreover, the more deeply these brain parts slept, the more the subject improved in performing the task the next day.

My favorite new finding suggests that sleep may not just rest our brains or reinforce what we know, but may help to create fresh insight. A team led by Ullrich Wagner of the University of Lübeck in Germany offered experimental evidence that sleep not only consolidates recent memories; by changing the way memories are structured in the brain, it may also foster breakthrough thinking and creative solutions to intractable problems.

In 2004, Wagner's team decided to investigate the creativity theory by testing subjects on a challenging task of logic. The subjects were asked to transform a series of strings of eight digits into new strings using two rules about pairing the digits, and then to deduce as quickly as possible the last number in the new sequence. They were not told of a hidden third rule, which was a quick shortcut to the answer. The subjects were trained in the task, tested, then given an eight-hour break before being retested. One group slept during the break; another remained awake. Of those who slept, 60 percent spotted the shortcut—more than twice as many as those who stayed awake. The use of the hidden rule could not have stemmed from practice, Wagner concluded, but must have arisen during sleep, when elements of the task learned in the training were rearranged. During sleep, as our brain shifts our memories from fresh to permanent status, it reorganizes them; this memory shuffling facilitates new insights.

So, says Wagner, the best way to deal with a problem may be to mull it over before going to bed and then just sleep on it.

The annals of literature hold numerous tales of insight emerging during or just after sleep. The poem "Kubla Khan" is said to have come to Coleridge in a dream. Robert Louis Stevenson claimed to have dreamed pivotal scenes in his novella *The Strange Case of Dr. Jekyll and Mr. Hyde.* "I had long been trying to write a story . . . on that strong sense of man's double being," he wrote. "For two days I went about racking my brains for a plot of any sort; and on the second night I dreamed the scene at the window, and a scene afterward split in two, in which Hyde, pursued for some crime, took the powder and underwent the change in the presence of his pursuers."

Science also holds stories of discoveries arising from somnolent states. Dmitri Mendeleev dreamed of a periodic table where all of the elements fell into their proper places. Breakthrough dreams on two consecutive nights aided Otto Loewi in envisaging the design for experiments that would reveal the chemical transmission of nerve impulses. One night the great chemist Friedrich August Kekulé was brooding over the mysterious structure of aromatic compounds such as benzene, which occur in fragrant oils and spices. He turned his chair to the fire and dozed. "The atoms were gamboling before my eyes," he wrote. "Long rows . . . all twining and twisting in snake-like motion. But look! What was that? One of the snakes had seized hold of its own tail, and the form whirled mockingly before my eyes."

Here, in the dream of a ring, was the solution to the structure of benzene.

Some would dispute these stories of discovery through sleep or dreams, arguing that the eureka! moments originated not in sleep but in that drowsy state of *dormiveglia,* when the conscious mind is still at work—or in Coleridge's case, in an opium-induced stupor. But the possibility that dreams might promote creative thought makes sense to me. I think back to that metaphoric dream I had a decade ago, of plunging headfirst into a bed of mud, which resolved my debate over medical school. Sometimes, in the face of a difficult decision, rational thinking, itemizing pros and cons, just doesn't work. Sleeping on it does, perhaps by creating a kind of practice run of various scenarios and testing our emotional response to them. Or perhaps by tying together unlike elements that we would never think to link while awake. Or perhaps just by giving us rest from our rational mind.

13

HOUR OF THE WOLF

•

IT'S 2 A.M., long past lights-out, but you murmur and toss, inching through darkness toward a night of no Morpheus. Who else is awake at this ungodly hour? Truck drivers, oil riggers, hospital doctors, pilots and air-traffic controllers, bakers, musicians, late-night partygoers heading home, and many, many of the elderly.

Aging sabotages both sleep and circadian rhythms. When William Dement and Mary Carskadon studied the sleep of healthy men and women ranging in age from sixty-five to eighty-eight, they found that most experienced frequent "microarousals"—the opposite of microsleeps. These brief awakenings may last only a few seconds, but they can occur between two hundred and a thousand times a night, badly disrupting deep sleep. This loss of deep sleep actually begins in midlife. Between the ages of thirty-six and fifty, less than 4 percent of our sleep time is of the deep variety, roughly a fifth of what we enjoyed in early adulthood.

To make things worse, aging may also derange the amplitude and the stability of our circadian rhythms. Some evidence suggests that in the elderly, the peaks of hormones such as melatonin and cortisol, as well as body temperature and other functions, are not as high, and the dips not as low. Older people often have extreme lark-like rhythms; their body temperature troughs well before dawn, for example, and they fall asleep and wake up much earlier than younger folks.

Circadian biologists are still trying to fathom what causes these shifts. They may in part be rooted in age-related changes in the eye —the yellowing of lenses, for instance, which blocks some of the light necessary to properly set circadian rhythms—or, possibly, changes in the SCN, the master clock. Scientists know that normal aging doesn't change the size of the SCN or the number of cells it contains. But at least one new study, by Gene Block and his colleagues at the University of Virginia, suggests that aging disrupts the functioning of SCN cells, especially their ability to synchronize the clocks in tissues throughout the body.

To learn how aging might affect the body's clock genes, the team studied old rats with clock genes genetically modified to carry a flashing luciferase gene, the kind that "blinks" in rhythm with the expression of the target gene. They found that the rhythms in the rats' SCN cells were normal, but the rhythms of cells in some of their outlying tissues were phase-advanced or even absent. Because the rhythms in these peripheral cells could be restored by applying a chemical, the team surmised that the problem might lie not in the cells' clocks themselves but in the failure of the SCN to send them appropriate signals. Perhaps, then, this missed communication between the "grandfather" clock and the tiny peripheral tickers accounts for the many old souls prowling their dimly lit kitchens in the dead of night.

In other cultures, such as the !Kung of Botswana, the Efe of the Congo, or the Gebusi of New Guinea, anyone at all might be awake now. In traditional, non-Western societies, social activity and frequent interruptions are often embedded in a night's sleep, says Carol Worthman, an anthropologist at Emory University. When Worthman conducted the first study of sleep patterns across a wide variety of traditional cultures, she discovered that the Western model of a habitual bedtime and a single spell of solitary sleep is rare indeed. Among the Efe, says Worthman, virtually no one sleeps alone, and "one may routinely find two adults, a baby, another child, a grandparent, and perhaps a visitor sleeping together in a small space." Arousals are common, with the movement and noises of others and the traffic of staggered bedtimes and trips to urinate. The Efe often go to sleep and then get up later because they hear something interesting going on—a conversation or music—and want to join in. Someone may wake up at any hour of the night and begin to hum or play the thumb piano or start a dance. The !Kung frequently spend the wee hours of the morning in lively dialogue, using nighttime chat to entertain themselves, debate, resolve conflicts and disputes, and work through troubled relationships. Gebusi men hold all-night séances and other social activities.

There was a time when deep night would have seen most Westerners, too, up and about or in a state of semiwakefulness. Not much was known about past patterns of sleep or nighttime activity in Europe until A. Roger Ekirch, a professor of history at Virginia Polytechnic Institute, brilliantly remedied the ignorance. In a study of historical records from 1300 to 1800, Ekirch found numerous references suggesting that most Europeans broke their rest into two phases: a "first sleep" and a "second," or "morning," sleep. (Some early medical books recommended lying on the right side during the first sleep and on the left during the second, to ease digestion and maximize comfort.) The two sleeps were usually bridged by an hour or more of quiet wakefulness. During this interlude, people often rose and moved about or stayed in bed to converse, pray, make love, reflect on the dreams that preceded their awakening, or just let their minds drift in a semiconscious state.

The habit of sleeping in bouts is typical of many mammals. Is it a more natural pattern for our nights, perhaps dating from our deep past? Did our prehistoric ancestors, too, enjoy a midnight interlude, the dark antipode of a midday siesta?

Thomas Wehr of the National Institute of Mental Health once devised an experiment that hinted at an answer. To mimic ancient winter nighttime conditions, he asked volunteers to spend fourteen hours a day in darkness (6 P.M. to 8 A.M.), without any artificial light, for a period of one month. In the first few weeks, the subjects slept in one long, consolidated stretch of up to eleven hours—perhaps in catchup from sleep deprivation. But eventually they fell into a pattern of two distinct periods, sleeping for four hours, from 8 P.M. to midnight, awakening from REM sleep and staying awake for a couple of hours in quiet, "nonanxious" rest, then falling back asleep at 2 A.M. for another four hours, until waking at 6 to start the day.

Wehr measured his subjects' temperature, hormones, secretion of melatonin, and EEG patterns and found that the chemistry of their nights differed from the norm, with higher levels of melatonin and sleep-related growth-hormone secretion throughout the night. The period of rest also had its own unique chemistry, with a dramatic rise in levels of prolactin, that hormone involved in lactation and (in chickens) brooding. This distinctive endocrinological state may have promoted self-reflection and a kind of quiet meditation, says Wehr. And the two-phase sleep pattern—waking from dream sleep into quiescent rest—may have provided a means of access to dreams that we've lost today. "It is tempting to speculate," Wehr writes, "that in prehistoric times this arrangement provided a channel of communication between dreams and waking life that has gradually been closed off as humans have compressed and consolidated their sleep. If so, then this alteration might provide a physiological explanation for the observation that modern humans seem to have lost touch with the wellspring of myths and fantasies."

Wehr suspects that our natural pattern is indeed the two-phase variety—at least during the long nights of winter—and that our current habit of consolidating sleep into a single bout is an artifact of contemporary life. "Modern humans no longer realize that they are capable of experiencing a range of alternative modes that may have once occurred on a seasonal basis in prehistoric times," he says, "but now lie dormant in their physiology." We have become clamped in a perpetual long-day/short-night pattern—and the squeeze is growing tighter.

You switch on your bedside lamp, one of the thousands of possible sources of artificial light by which we shorten our nights and deny ourselves access to dream life. Thomas Edison, more than anyone, put darkness and reverie in retreat.

For tens of thousands of years, nightfall was the signal for humans to sleep; sunrise, the signal to awaken. Sunlight was the only light source for recalibrating our internal clocks, setting them to the cycles of the day and also the season. After that came wood fires and lamps that burned animal fat and pitch and petroleum, which supplied light enough to see by but not sufficient to reset our biological clocks. Then came the invention of the incandescent light bulb in 1879, and the rapid spread of artificial light. Suddenly our species seemed liberated from the fetters of the solar cycle, able to pretend that every night was a midsummer's night. However, because our inner clocks still adhere to an ancient light-dark schedule, there are costs associated with this round-the-clock illumination—the magnitude of which we're only now discovering.

Our body clocks need darkness as desperately as they need light. In 2005, scientists at Vanderbilt University showed that constant light desynchronizes the firing of neurons that make up the SCN. In turning on lamps and lights after the sun has set, we unintentionally reset our biological clocks. Exposure to even low light levels of, say, 100 lux—similar to the ambient lighting of offices and living rooms—can affect the phase of our rhythms. Charles Czeisler's team has found that during the first hours of biological night, our circadian pacemakers are especially vulnerable. Light exposure in the late evening delays the phase of our clock, so that it acts as if sunrise comes later. Exposure in the early morning advances the clock, so it expects sunrise earlier. Light at night suppresses the production of the hormone melatonin. Even brief exposure to light in the middle of the night radically reduces the activity of an enzyme necessary to make melatonin.

Ours is the only species that lights up its biological night, that overrides its own rhythms, crosses time zones, works and sleeps at times that run counter to its internal clocks. We ignore what our clocks remember at our own peril.

Take transmeridian travel. Not long ago, I sat at a table in a small village in China with a number of local dignitaries and scientists. Before us was an array of beautiful and exotic foods, including bowls of *ni do*, bird's nest soup. The soup, as I understood, is made from the nests of swiftlets, surprisingly sturdy creations woven from strands of gummy saliva, which are simmered in chicken broth. I knew I had to try the expensive delicacy, but with spoon raised, I hesitated. It's not that I objected to feasting on gelatinous strands from a bird's salivary glands. I prided myself on being an adventurous eater, ready to sample all sorts of authentic foods. But my stomach would have none of this dish—or any other. I had arrived in China only a day earlier, and my insides felt as if they were still back in Virginia. In fact, a scientist friend later told me, they were.

When you fly halfway around the world, it is said, your soul takes about three days longer to get there. So does your stomach. Indeed, you may seem to arrive in one piece, says Michael Menaker, but clockwise, different parts of your body follow only slowly. For each time zone crossed, it can take up to a day for your systems to fully adjust to the new time. Two-thirds of time-zone travelers report the symptoms of jet lag—that muzzy sluggishness and upset stomach, daytime fatigue, trouble going to sleep at night (after eastward flight) or waking up too early (after westward flight), lapse of memory and alertness, and loss of appetite, to name a few. Travelers are often awoken in the middle of the night by the surge of hormones that signal morning. Symptoms are generally worse going from west to east, perhaps because it's easier for the body to adjust to a longer day than to a compressed one.

Menaker suspects that the discomfort and malaise of jet lag arise from a loss of synchrony between the body's master clock and its peripheral clocks, and also among these outlying clocks, as they attempt to catch up in a new time zone, each at its own pace. In one study, Menaker and his colleagues used genetically modified rats to monitor the impact of time shifts on the circadian rhythms of different organs. The results suggested that the master clock in the brain's SCN, which oversees our big rhythms, such as body temperature, gets back on track within about a day; the peripheral clocks in our tissues, however—those in the lungs, muscles, and liver, for instance—may take a week or more to catch up. When the brain signals the muscles to exercise, the muscles may not respond well, as their clock still has them in deep sleep. Likewise, my brain may say it's time for *ni do* in China, but to my Virginia-moored liver, it's still the middle of the night.

This lag time in clock modification is essential in normal life. If our inner clocks shifted instantaneously with sudden changes in light, they would be spinning forward and backward each time we entered or exited a dark room. The system is designed in such a way as to adjust easily to small, gradual changes in patterns of light and dark, such as seasonal changes in day length. "But transmeridian flight is an unnatural event for which the body is not prepared," says Menaker. "Crossing time zones causes large and abrupt shifts in the light cycle, which severely disorganize the system."

Leaping time zones occasionally is one thing. Doing it often is quite another. Kwangwook Cho of the University of Bristol was inspired by his own jet lag symptoms—disorientation and memory lapses—to look into the effects of frequent transmeridian travel. In a study of twenty flight attendants working for international airlines, Cho found that five years of long-haul travel caused memory problems and cognitive impairment. Further probing with saliva samples and brain scans revealed the possible cause: Attendants who flew over more than seven time zones with fewer than five days to recover between multizone flights showed boosted levels of the stress hormone cortisol. When the body is constantly subjected to the bewildering signals of light and dark that come with long-distance travel, it grows confused about whether it's night or day and keeps making cortisol around the clock. As we know from studies on chronic stress, cortisol in high concentrations damages brain cells. Sure enough, the brain scans revealed shrinkage of the attendants' temporal lobe, including the hippocampus, the part of the brain so essential to learning and memory.

The swish of distant traffic sounds on a nearby highway. Who else is up in the black of night? A couple of hundred years ago, only night watchmen, gatekeepers, and perhaps the occasional cook worked through the hours of darkness. Now, some 15 percent of the U.S. workforce is made up of people who labor into the night, controlling air traffic, driving trucks, and running hospitals, fire and police stations, factories, and nuclear power stations—work that puts them radically out of phase with natural time cues.

To flesh out the effects of nighttime labor on the body, Josephine Arendt of the Center for Chronobiology in Surrey, England, has conducted extensive field studies of workers on oil rigs in the North Sea. The work on these rigs is challenging and dangerous, she says. "To get a job there, you have to pass a test where you're hung upside down in the water by helicopter. If you can get yourself out of this predicament, you can work on the rigs—though you're expected to work difficult schedules."

Studying the laborers is no piece of cake, either. "Getting forty-five rig workers to collect urine every four hours for fourteen days is a feat," says Arendt. "Rigs make about three million dollars a day, so interruptions of work to pee for circadian studies are not particularly welcome." Nevertheless, Arendt got her results. She compared laborers working two different shift schedules for two weeks: one a simple twelve-hour shift, with workers on either night shift or day shift; the other a "swing" rotation of seven night shifts followed by seven day shifts.

"The swing schedule was the worst," says Arendt. The urine tests from workers on this schedule revealed that their melatonin levels never synchronized with their new hours, so they had difficulty sleeping. This is true for many shift workers, who try to sleep in phases at odds with their circadian cycles, when melatonin is declining and body temperature rising. Their sleep is disjointed, and they wake up as fatigued as ever. "It's important to go to sleep at the right circadian phase," explains Arendt. "If you go to sleep during your biological day, after your temperature nadir, sleep is of poor quality." Studies suggest that shift work reduces sleep time by an average of three to four hours a night.

Arendt found signs of other serious long-term health effects. When workers on either schedule had to eat their meals late at night, their blood showed abnormally high levels of fatty acids linked with heart disease and also a reduced tolerance for glucose, a risk factor for diabetes and other metabolic disorders.

Shift work probably has the same desynchronizing effect on the circadian system as chronic jet lag: Shift workers' out-of-kilter clocks affect memory, cognition, and various body systems, causing high cholesterol, high blood pressure, mood disorders, infertility, and a higher risk of heart attack and cancer.

Research on the 78,500 women in the Nurses' Health Study revealed that nurses who had worked a graveyard shift for ten years had a 60 percent greater risk of breast cancer and a higher risk of colon cancer compared with those who did not work at night. A few years later, a Japanese study of more than 14,000 men, showed that workers who swing between day and night shifts had triple the normal rate of prostate cancer. And in experiments where researchers have tampered with the circadian rhythms of mice to mimic shift-work conditions, tumor growth accelerates.

What might account for the link between shift work and cancer?

Some scientists suspect that the answer may lie deep in our genes. The disrupted rhythms caused by shift work may dramatically change the expression of clock genes, which in turn may affect "downstream" genes that control growth. In a 2006 study, William Hrushesky and his colleague Patricia Wood showed that the body's clock genes "gate," or regulate, the enzymes that control DNA synthesis, cell division, and blood vessel formation both in normal tissue in the gut and bone marrow and, at different times, in cancerous tissue.

Artificial light has also been implicated in the circadian disruption–cancer connection. For years science has known that exposure to nighttime illumination curbs the body's normal production of melatonin. And animal studies have shown that suppressed melatonin release boosts the growth of cancers. But the first strong experimental evidence of a tie between the two came in 2005.

A team of researchers drew blood from a group of twelve women three times over the course of twenty-four hours: during the day, at night, and again after exposure to bright light at night. Then the team injected the different blood samples into a human-breast tumor that had been implanted in a rat. The results showed that the tumor grew most rapidly when exposed to the blood drawn from the daytime sample and from the nighttime illumination sample. Both samples contained little melatonin. The study, say the researchers, strongly suggests that exposure to artificial light at night dampens melatonin production, which spurs tumor growth. Hence, possibly, the increased breast cancer risk in female night-shift workers.

Asking workers to labor through the night and to work different night schedules creates a major health hazard, not just for the individual worker, says Arendt, but for society at large. When workers are disoriented by circadian dysfunction and fatigued from lost sleep, accidents happen. The explosion at the Union Carbide plant in Bhopal, India, in 1984, which killed thousands of people, occurred just after midnight. The 1979 crisis at the Three Mile Island nuclear plant in Pennsylvania began at 4 A.M., when workers who had just moved from a day shift to a night shift failed to notice a stuck valve. And the world's worst nuclear accident, at the Chernobyl plant in Ukraine, in 1986, began at 1:23 A.M. as a result of a series of errors made by night-shift operators.

If I really wanted company this time of night, I know where I'd go. The lamps are burning brightly at the teaching hospital down the street, where physicians in training are working up to thirty hours straight and as many as eighty hours per week.

Here's another example of modern life pushing the circadian envelope. The tradition of long work hours for medical interns in this country is the legacy of William Steward Halsted, a brilliant surgeon who worked around the turn of the twentieth century at Johns Hopkins Hospital in Baltimore. In medical circles, it's well known that Halsted promoted the idea that young physicians should live at the hospital and work around the clock, the better to be exposed to as many patients as possible. What most people don't know is that Halsted was addicted to cocaine. Today his system of "heroic" schedules remains the hallmark of medical education, despite mounting evidence of its risks.

Right about now, with no sleep in the last twenty-four hours, how well would you do making a critical decision about diagnosis, dosage, possible danger to a patient? How would you feel about being treated by a young doctor in a similar position?

Despite all the research showing that sleep deprivation impairs cognitive performance, until recently there have been few studies measuring its effects on medical errors. In 2004 a team of scientists at the Harvard Work Hours, Health, and Safety Study reported that medical interns working thirty-hour shifts suffered twice as many failures of attention while working at night as interns whose work was limited to sixteen consecutive hours. Moreover, the thirty-hour group made significantly more serious medical errors and five times as many major diagnostic errors. "When people have been awake for 17 to 19 hours, their performance is equivalent to someone with a blood alcohol level of .05 percent," explains team member Charles Czeisler. When they're awake for twenty-four hours, it's .10 percent. "The risk of making a mistake after working for 24 hours is so great," he says, "that sleep experts and legislators in Massachusetts have suggested it may be ethically imperative for hospitals to notify patients if the doctor treating them has been awake for 22 of the past 24 hours."

The safety risks of interns' long work hours are not confined to hospital patients; the interns themselves are endangered. A 2006 study by the Harvard group found that interns working these marathon shifts had a 61 percent increased risk of stabbing themselves with a needle or scalpel while working, thereby exposing themselves to the risk of contracting hepatitis, HIV, and other blood-borne illnesses. A lapse in concentration and fatigue were the most commonly reported factors contributing to the accidents.

Overworked interns may also be a hazard to themselves and to others when they're on the road home from their shifts. Studies of people who sleep only five or six hours a night on a regular basis (the average for most interns) have found that their median reaction time triples. "That means that when interns are driving home from an extended shift, if a kid dodges in front of their car," says Czeisler, "it will take them three times longer to move the steering wheel or put their foot on the brake." In 2005, Czeisler's group reported that interns working an extended shift of more than twenty-four hours had double the risk of having a car crash while driving home, and five times the risk of a near miss.

"In the face of this evidence, it might be reasonable to ask what the medical profession is doing to address the problem," suggests Chris Landrigan, a member of the Harvard team. In 2003, the profession implemented national work-hour limits for physicians in training. "But the rules still allow interns and residents to work as many as thirty hours straight—far beyond the limit considered acceptable in other safety-sensitive industries," Landrigan says. Pilots, truckers, and nuclear plant workers, for example, are all limited to eight to twelve consecutive hours of work. Moreover, a study conducted by Landrigan in 2006 reported that 84 percent of interns don't comply with the limits. "Limiting residents' work hours undoubtedly represents a cultural and financial challenge within medicine," says Landrigan, "but it's one that has been successfully tackled in other countries: in the United Kingdom and New Zealand, for instance, where the consecutive work hours of physicians in training are limited to 13 and 16, respectively."

Now imagine that the young American doctor treating you had a drug that could defeat drowsiness, a pick-me-up pill popped in the morning that promised two full days of perpetual wakefulness. There are, in fact, such "lifestyle" drugs, designed to murder drowsiness and improve cognitive performance, among them, modafinil and CX717, described as "unique wake-promoting agents" with "unknown specific mechanisms of action." CX717 is under study as a wakefulness promoter for soldiers in combat. Called eugeroics (from the Greek, meaning "good arousal"), these drugs seem to have few of the drawbacks of other stimulants: the jitters, the risk of addiction, the post-pill crash. Despite their benefits, however, they're not always entirely effective. According to a 2005 study by Czeisler's team, for instance, some people who take modafinil to get through a night of work may still have excessive sleepiness and impaired performance.

How safe is it for physicians or other safety-sensitive workers to operate under the assumption that they're alert because they took a pill that's supposed to make them so? More and more, doctors are prescribing the drug for shift workers, pilots, truck drivers, and fellow doctors to sustain alertness. And just over the pharmacological horizon is another wave of drugs, these offering a condensed dose of sleep purportedly more restorative than the natural kind, reducing need for the real time-out.

No one knows the long-term effects of sabotaging natural sleep and tampering with the body's timepieces. Just how far will we go to oblige a twenty-four-hour society?

Have you dozed? It's sometime between 3 and 4 A.M., the peak hour for night-work errors, for auto and truck crashes, for congestive heart failure and gastric ulcer crisis, for sudden infant death syndrome and bone breakdown, for migraine headaches and asthma attacks. In a few hours will come the peak hour for dying of any cause—perhaps because of the rise in blood pressure and boost in cortisol occurring now, in anticipation of waking time. It seems odd that we should more often finish life not toward the close of day but at its start, as if to deny that death is an ending. But that's the way of the body, all paradox, surprise, contradiction.

This, the hour of the wolf, is also when body temperature falls to its nadir, and spirits too, when fears loom large, and regrets, and misgivings. "In the real dark night of the soul it is always three o'clock in the morning," wrote F. Scott Fitzgerald. Now, when sleep should be sweet and deep, the wakeful mind worries things kept at bay by daylight and distraction: remorse over a misspoken word or a love not bestowed, mounting debt, the frenetic pace of life, the creeping decrepitude of aging.

Look up the word "time" in *Webster's Third New International Dictionary* and you'll find that it takes up more space than "life" or "love," more than "God" or "truth." And that's not counting the myriad compounds and time-bound expressions, time-out, timeworn, time flying or racing, time lost, seconds split, time relative but always relentless. "The gods confound the man who first found out how to distinguish hours," wrote the Roman playwright Plautus. "Who in this place set up a sundial, to cut and hack my days so wretchedly into small pieces!"

Since Plautus, we have become even more hopelessly time-minded, mincing our days into smaller and smaller portions. In the past half century, we have increased by orders of magnitude the precision with which we measure time. Quartz clocks are accurate to one second a month; the finest cesium atomic clocks, to one second in thirty million years. In 2005, scientists devised an "optical lattice" clock that is a thousand times more accurate than even the cesium clock; it uses the element strontium, which "ticks" at 429,228,004,229,952 times each second.

As if all of this precise measurement would make time flow as we wish—swifter during acute pain, for instance, or slower as we age.

Vladimir Nabokov once said that the first creatures on earth to become aware of time were also the first creatures to smile. I don't know. When digital clocks first came into fashion, I mourned the loss of the circular clock face, with its hands mimicking the cyclical sweep of shadow over a sundial. Perhaps this was simple nostalgia for youth, when I could run the perimeter of the playground or smear circles of finger paint, red, blue, and yellow, round and round into slick loops of brown, and believe at the end that I was precisely the same age I had been at the start.

Up to a certain age, time feels, well, cyclical. Dawn to dawn, one day rounding into the next, every end a beginning, and so on until you're forty or fifty, when the problem in life, as Virginia Woolf wrote, is "how to grasp it tighter and tighter to you, so quick it seems to slip, and so infinitely desirable is it." Suddenly the twenty-eight thousand days in a long human life seem cruelly brief and inadequate. Suddenly you're well on your way to old, to tooth rot, sagging chin and gimpy knee, senility.

These days, time seems a fleet arrow indeed, which makes all the more comforting a surprising secret gleaned about the clocks that run our bodies: In a way, they defy linear time.

What may lie at the heart of all the body's clocks is an ingenious self-winding mechanism that enables a cell to tell time. A set of genes interact in tight negative-feedback loops to produce the oscillation, or tick-tock, in their own expression. Some of these genes make daylight proteins, which accumulate during the day. When they reach a peak level in the evening, they shut off the biochemical activity that leads to their own production. The result is a robust, self-sustaining loop that cycles continuously over twenty-four hours.

Imagine if we could feel the rotation of these intimate circles that drive our bodies. Perhaps this would temper our habit of ticking off digitized minutes and linear hours; perhaps it would restore our childlike experience of time as a revolving ring made of the smaller circumferences of a day. At the very least it might nurture more respect for our natural cyclical rhythms.

There's a thin sliver of moon in the sky. I lie for a moment in the dark and think about the body's little loops, its inner lark or owl and helpful microbial handmaidens, its exquisite senses and love of natural light, its need for sleep. Though we still know little about many aspects of bodily existence, the gaps in our knowledge are narrowing.

The body is like an Antarctica, a continent being opened up, mapped, even transformed. With new understanding come new and useful tools for making the best of our own strange and temporary vessels.

With the first pale hints of light on the horizon, sleep finally takes possession, sweet and irresistible, subduing thoughts and senses. Whatever its purpose, thank goodness each day dies with it. There are some who would choose to run uninterrupted around the clock. But I can't imagine pressing on relentlessly through day and night with mind, body, spirit in a single state, can't imagine denying myself the possibility of a fresh start.

ACKNOWLEDGMENTS

NOTES   INDEX

Acknowledgments

I WOULD LIKE TO THANK three scientists who helped me in various ways and who have devoted their careers to the study of time in the body: Mike Menaker of the Center for Biological Timing at the University of Virginia, who read portions of the manuscript at various stages and offered helpful suggestions; William Hrushesky of the University of South Carolina, whose excellent 1994 article on timing in health and disease in *The Sciences* first introduced me to the topic of chronobiology, and who later suggested readings on chronotherapy; and Michael Smolensky of the University of Texas, whose book, written with Lynne Lamberg, *The Body Clock Guide to Better Health* (Henry Holt, 2000), was a welcome and authoritative guide to the role of the biological clock in health and daily life.

I cannot begin to adequately thank these scientists and the many others who helped me prepare this book. For their generous and kindly guidance, suggestions for additional reading, and corrections of my manuscript, I am especially indebted to Josephine Arendt of the University of Surrey; Paul Breslin of the Monell Chemical Senses Center; David E. Cummings of the University of Washington; William C. Dement of Stanford University School of Medicine, author, with Christopher Vaughan of the outstanding book *The Promise of Sleep* (Random House, 1999); A. Roger Ekirch of Virginia Tech, who offered not only helpful suggestions and encouragement but also inspiration in the form of his masterly work on night, *At Day's Close: Night in Times Past* (Norton, 2005); Helen Fisher of Rutgers University; Jeffrey

Gordon of Washington University; Jay A. Gottfried of Northwestern University; Carla Green of the University of Virginia; Jack Gwaltney of the University of Virginia, who fielded my many questions about the common cold; H. Craig Heller of Stanford University; Richard Ivry of the University of California at Berkeley; Eric Kandel of Columbia University; Christof Koch of the California Institute of Technology; Art Kramer of the University of Illinois; Christopher Landrigan of the Harvard Work Hours, Health, and Safety Study at Harvard Medical School; Joseph LeDoux of New York University; Daniel E. Lieberman of Harvard University; Bruce McEwen of Rockefeller University, whose book *The End of Stress as We Know It* (Dana Press, 2002) was a superb tour of the subject; Janet Metcalfe of Columbia University; Thomas Reilly of Liverpool John Moores University; Craig Roberts of the University of Liverpool; Till Roenneberg of the Centre for Chronobiology at Ludwig-Maximilians-University in Munich; Mel Rosenberg of Tel Aviv University; Timothy Salthouse of the University of Virginia; Sally and Bennett Shaywitz of the Yale School of Medicine; Jerome Siegel of the University of California at Los Angeles; Ullrich Wagner of the University of Lübeck; and Charles Wysocki of the Monell Chemical Senses Center.

I am also deeply grateful to the following scientists, who read sections of the manuscript and offered useful comments: Paul Bach-y-Rita of the University of Wisconsin, James Blumenthal of Duke University, Jan Born of the University of Lübeck, Richard A. Bowen of Colorado State College, Arthur Burnett of Johns Hopkins School of Medicine, William Carlezon of Harvard University, Mary Carskadon of Brown University, Priscilla Clarkson of the University of Massachusetts, Richard Cytowič, Angelo Del Parigi of Pfizer Global R & D, Scott Diamond of the University of Pennsylvania, Brad Duchaine and Henrik Ehrsson of University College London, Jeffrey Flier of Harvard Medical School, Kevin Foster of Harvard University, Lynn Hasher of the University of Toronto, J. Owen Hendley of the University of Virginia, J. Allan Hobson of Harvard University, Gert Holstege of the University of Groningen, Jim Hudspeth of Rockefeller University, Laura Juliano of American University, Philip Kilner of the Imperial College of London, Kristen Knutson of the University of Chicago, Barry Komisaruk of Rutgers University, Peretz Lavie of Technion–Israel Institute of Technology, Peter Lucas of George Washington University, Sara Mednick of the Salk Institute for Biological Studies, David Meyer of the University of Michigan, Michael Miller of the University of Maryland Medical Center, Tore Nielsen of the University of Montreal, Tim Noakes of the University of Cape Town, Charles P. O'Brien of the University of Pennsylvania, Håkan Olausson of Sahlgrenska University Hospital, Steven Platek of the University of Liverpool, George Preti of the Monell Chemical Senses Center, Eric Ravussin of the Pennington Biomedical Research Center, Naftali Raz of Wayne State University, Allan Rechtshaffen, Marianne Regard of University Hospital Zürich, Michael Sayette of the University of Pittsburgh, Dee Silverthorn of the University of Texas, Dana Small of Yale University Medical School, Esther Sternberg of the National Institutes of Health, D. Michael Stoddart of the University of Tasmania, Henning Wackerhage of the University of Aberdeen, Peter Weyand of Rice University, Carol Worthman of Emory University, and Shawn Youngstedt of the Arnold School of Public Health at the University of South Carolina.

Errors may yet lurk in these pages, but it is thanks to those I've noted in the paragraphs above that there aren't many more.

In the course of the four years it took to write this book, I was fortunate enough to receive a fellowship in nonfiction from the National Endowment for the Arts. I could not have finished the book without this support. My appreciation goes to the late Cliff Becker for his interest in my project and, especially, for his passionate championing of literature.

For many different kinds of help, deep thanks to my close friend and collaborator Miriam Nelson, who offered in spades the kind of expertise, support, and enthusiasm treasured by her large circle of friends. I'm also very grateful to Francesca Conte, Harri Wasch, and Heather Sellers, who shared with me tales from their lives, and to my friend Dan O'Neill, who read an early draft of the manuscript and made many intelligent and helpful suggestions. Thanks also to my fine editors at the *Yale Alumni Magazine,* Kathrin Lassila and Bruce Fellman, and at *National Geographic:* Lynn Addison, Oliver Payne, Jennifer Reek, and Caroline White.

I'm grateful to Laurence Cooper for his careful editing of the manuscript, to Will Vincent for his competent and cheerful help with the publishing process, and to Martha Kennedy for her inspired book title suggestion and jacket design. Special thanks to Janet Silver for her patience and generous support and to Melanie Jackson, always a font of wisdom and good judgment. To Amanda Cook I will say only this: You're the sort of editor every writer dreams of—smart, talented, articulate, and funny to boot. I can't thank you enough for your help in shaping, smoothing, and polishing this book.

Finally, my deepest gratitude goes to my dear girls, Zoë and Nell, and to my husband, Karl, for his love and for his willingness to take on, with characteristic courage and fortitude, the outside world while I was engrossed in the inside one.

Notes

page **PROLOGUE**

xiii *we have learned that the human body:* P. B. Eckburg et al., "Diversity of the human intestinal microbial flora," *Science* 308, 1635–38 (2005).
"*timing is everything*": "Timing is everything," *Nature* 425, 885 (2003).
"*Nature is nowhere accustomed*": Thomas Willis quoted in Oliver Sacks, "To see and not see," *The New Yorker*, May 10, 1993, 59.

xiv "*Our body is like a clock*": Robert Burton, *The Anatomy of Melancholy.* Available at www.psyplexus.com/burton/7.htm.

xv "*I should not talk so much*": Henry David Thoreau, "Economy," in *Walden and Other Writings of Henry David Thoreau* (New York: Modern Library, 1992), 3.

1. AROUSAL

3 *Some people claim that subtle aural trigger:* "Beating the bell," *New Scientist,* letters by Jim Field and Radko Osredkar, May 14, 2005.
Scientists at Brown University documented: M. A. Carskadon and R. S. Herz, "Minimal olfactory perception during sleep: why odor alarms will not work for humans," *Sleep* 27:3, 402–5 (2004).

4 *When Peretz Lavie, a sleep researcher:* Peretz Lavie et al., "It's time, you must wake up now," *Perceptual and Motor Skills* 49, 447–50 (1979).
Another study showed: Jan Born, "Timing the end of nocturnal sleep," *Nature* 397, 29–30 (1999).
In a nation that averages: Till Roenneberg et al., "Life between clocks: daily temporal patterns of human chronotypes," *Journal of Biological Rhythms* 18:1, 80–90 (2003).
Unfortunately, the short bouts: Edward Stepanski, Rush University Medi-

cal Center, Chicago, quoted in Martica Heaner, "Snooze alarm takes its toll on nation," *New York Times,* October 12, 2004, D8.

4   *If the snoring sinner ignored:* "An alarming bed," *Scientific American,* October 1955, reprinted in *Scientific American,* October 2005, 16.
Only slightly more humane: http://alumni.media.mit.edu/~nanda/proj ects/clocky.html.

5   *"The brain just doesn't go from 0 to 60":* Quote and anecdote about the U.S. Air Force are from Charles Czeisler, "Sleep: what happens when doctors do without it," Medical Center Hour, University of Virginia School of Medicine, Charlottesville, March 1, 2006.
When a team of scientists: K. W. Wright et al., "Effects of sleep inertia on cognition," *Journal of the American Medical Association* 295:2, 163 (2006).
Lavie's team found that people: Lavie et al., "It's time, you must wake up now."

6   *To eliminate such rude awakenings:* See the SleepSmart Web site: www .axonlabs.com/pr_sleepsmart.html.
Whether you hop or drag: Roenneberg et al., "Life between clocks."
Such is the pattern, too, for the great geneticist: Jonathan Weiner, *Time, Love, Memory* (New York: Knopf, 1999), 190.
The "birds" differ dramatically: Michael Smolensky and Lynne Lamberg, *The Body Clock Guide to Better Health* (New York: Holt, 2000), 40–42.
Till Roenneberg, a chronobiologist: Roenneberg et al., "Life between clocks."

7   *You can assess your own status:* www.imp-muenchen.de/index.php?id=932.
Some time ago, a group of British researchers: C. Gale, "Larks and owls and health, wealth, and wisdom—sleep patterns, health, and mortality," *British Medical Journal,* December 19, 1998, E3 (col. 5).
Almost a decade ago, Hans Van Dongen: H.P.A. Van Dongen, "Inter- and intra-individual differences in circadian phase," Ph.D. thesis, Leiden University, Netherlands, ISBN 90-803851-2-3 (1998); H.P.A. Van Dongen and D. F. Dinges, "Circadian rhythms in fatigue, alertness, and performance," in M. H. Kryger et al., *Principles and Practice of Sleep Medicine,* 3rd ed. (Philadelphia: W. B. Saunders, 2000). See also J. F. Duffy et al., "Association of intrinsic circadian period with morningness-eveningness, usual wake time, and circadian phase," *Behavioral Neuroscience* 115:4, 895–99 (2001).
Although you might be able to overcome: Hans Van Dongen Q & A at www .upenn.edu/pennnews/current/2004/092304/cover.html, retrieved March 17, 2005.
"Time is the substance": Jorge Luis Borges, "A New Refutation of Time," *Labyrinths* (New York: Modern Library, 1983), 234.
To understand this, think back: The following derives from Ezio Rosato and Charlambos P. Kyriacou, "Origins of circadian rhythmicity," *Journal of Biological Rhythms* 17:6, 506–11 (2002); Russell Foster and Leon Kreitzman, *Rhythms of Life* (London: Profile Books, 2004), 157 f.

8   *"In this way . . . the circadian pacemaker"*: T. A. Wehr, "A 'clock for all seasons' in the human brain," in R. M. Buijs et al., eds., *Progress in Brain Research* 111 (1996).

So sensitive are these pacemakers: The following draws on Foster and Kreitzman, *Rhythms of Life*, 11.

special light-sensitive cells: M. S. Freedman et al., "Regulation of mammalian circadian behavior by non-rod, non-cone, ocular photoreceptors," *Science* 284, 502–4 (1999); D. M. Berson et al., "Phototransduction by retinal ganglion cells that set the circadian clock," *Science* 295, 1070–73 (2002); I. Provencio, "Photoreceptive net in the mammalian retina," *Nature* 415, 493 (2002); S. Hattar et al., "Melanopsin-containing retinal ganglion cells: architecture, projections, and intrinsic photosensitivity," *Science* 295, 1065–68 (2002); I. Provencio et al., "A novel human opsin in the inner retina," *Journal of Neuroscience* 20, 600–605 (2000); R. G. Foster, "Bright blue times," *Nature* 433, 698–99 (2005); Z. Melyan et al., "Addition of human melanopsin renders mammalian cells photoresponsive," *Nature* 433, 741–45 (2005); D. M. Dacey et al., "Melanopsin-expressing ganglion cells in primate retina signal colour and irradiance and project to the LGN," *Nature* 433, 749–51(2005).

9   *As Emerson wrote:* Ralph Waldo Emerson, "Circles," in *Essays and Poems* (London: Everyman Paperback Classics, 1992), 147.

the true average daily temperature: P. A. Mackowiak et al., "A critical appraisal of 98.6 degrees F, the upper limit of the normal body temperature, and other legacies of Carl Reinhold August Wunderlich," *Journal of the American Medical Association* 268, 1578–80 (1992).

Our knack for holding steady: The following information on homeostasis is from Foster and Kreitzman, *Rhythms of Life*, 53–54.

An intricate and diverse network: Catherine Rivier Laboratory Web site, www.salk.edu/LABS/pbl-cr/02_Research.html, retrieved March 11, 2006.

our set-points aren't set at all: The following discussion of circadian rhythms in body function comes from Wehr, "A 'clock for all seasons' in the human brain"; T. Reilly et al., *Biological Rhythms and Exercise* (New York: Oxford University Press, 1997), 50; Y. Watanabe et al., "Thousands of blood pressure and heart rate measurements at fixed clock hours may mislead," *Neuroendocrinology Letters* 24:5, 339–40 (2003); D. A. Conroy et al., "Daily rhythm of cerebral blood flow velocity," *Journal of Circadian Rhythms* 3:3, DOI: 10.1186/1740-3391-3-3 (2005); W.J.M. Hrushesky, "Timing is everything," *The Sciences*, July/August 1994, 32–37; John Palmer, *The Living Clock* (New York: Oxford University Press, 2002); Foster and Kreitzman, *Rhythms of Life*, 10–21.

10 *Some scientists even argue:* Foster and Kreitzman, *Rhythms of Life*, 71.

The rest of us can use: Foster and Kreitzman, *Rhythms of Life*, 11; Smolensky and Lamberg, *The Body Clock Guide to Better Health*, 5–12; Hrushesky, "Timing is everything."

So pervasive is the influence: J. Arendt, "Biological rhythms: the science of chronobiology," *Journal of the Royal College of Physicians of London* 32, 27–35 (1998).

11 *These clusters of . . . neurons:* P. L. Lowrey and J. S. Takahashi, "Mammalian circadian biology: elucidating genome-wide levels of temporal organization," *Annual Review of Genomics and Human Genetics* 5, 407–41 (2004).

In one 2004 study, researchers: S.-H. Yoo et al., "Period2: luciferase real-time reporting of circadian dynamics reveals persistent circadian oscillations in mouse peripheral tissues," *Proceedings of the National Academy of Sciences* 101, 5339–46 (2004).

Though the master clock: S. Yamazaki et al., "Resetting central and peripheral circadian oscillators in transgenic rats," *Science* 288, 682–85 (2000).

12 *Louis Ptáček and his colleagues:* C. R. Jones et al., "Familial advanced sleep-phase syndrome: a short-period circadian rhythm variant in humans," *Nature Medicine* 5:9, 1062 (1999); K. L. Toh et al., "An hPer2 phosphorylation site mutation in familial advanced sleep phase syndrome," *Science* 291, 1040–43 (2001).

British scientists have shown: S. Archer et al., "A length polymorphism in the circadian clock gene *Per3* is linked to delayed sleep phase syndrome and extreme diurnal preference," *Sleep* 26:4, 413–15 (2003).

A team of scientists gave: D. Katzenberg, "A clock polymorphism associated with human diurnal preference," *Sleep* 21:6, 568–76 (1998).

"It seems that our parents": C. M. Singer and A. J. Lewy, "Does our DNA determine when we sleep?," *Nature Medicine* 5, 983 (1999).

When Till Roenneberg studied: Till Roenneberg, "A marker for the end of adolescence," *Current Biology* 14:24, R1038–39 (2004).

13 *Research by Roenneberg suggests:* Till Roenneberg, personal communication, September 8, 2006; Roenneberg et al., "Life between clocks."

Bach loved coffee: S. M. Somani and P. Gupta, "Caffeine: a new look at an age-old drug," *International Journal of Clinical Pharmacology, Therapeutics, and Toxicology* 26, 521–33 (1988).

Two hundred years ago: Samuel Hahnemann, *Der Kaffee in seinen Wirkungen* (Leipzig, 1803), quoted in Bennett Alan Weinberg and Bonnie K. Bealer, *The World of Caffeine* (New York: Routledge, 2002), 119.

More than 80 percent: Jack James, *Understanding Caffeine* (Thousand Oaks, Calif.: Sage Publications, 1997); Laura Juliano, personal communication, October 2006.

14 *Members of the Achuar Jivaro tribe:* W. H. Lewis et al., "Ritualistic use of the holly *Ilex guayusa* by Amazonian Jivaro Indians," *Journal of Ethnopharmacology* 33:1–2, 25–30 (1991).

Czeisler and his team at Harvard: J. K. Wyatt et al., "Low-dose repeated caffeine administration for circadian-phase-dependent performance degradation during extended wakefulness," *Sleep* 27, 374–81 (2004). The study was designed to find the best strategy for boosting and sustaining alertness among workers who have to stay awake for long hours—for example, doctors and long-distance truck drivers.

Just why caffeine has: Jean-Marie Vaugeois, "Positive feedback from coffee," *Nature* 418, 734–36 (2002).

Within fifteen to twenty minutes: J. Blanchard and S.J.A. Sawers, "The absolute bioavailability of caffeine in man," *European Journal of Clinical Pharmacology* 24, 93–98 (1983).

Caffeine enhances alertness: J. W. Daly et al., "The role of adenosine receptors in the central action of caffeine," in B. S. Gupta and U. Gupta, eds., *Caffeine and Behavior: Current Views and Research Trends* (Boca Raton, Fla.: CRC Press, 1999), 1–16.

15   *Whether it actually perks up:* H.P.A. Van Dongen et al., "Caffeine eliminates psychomotor vigilance deficits from sleep inertia," *Sleep* 24:7, 813–19 (2001); L. M. Juliano and R. R. Griffiths, "A critical review of caffeine withdrawal: empirical validation of symptoms and signs, incidence, severity, and associated features," *Psychopharmacology* 176, 1–29 (2004).

In 2005, a team of Austrian scientists: F. Koppelstatter et al., "Influence of caffeine excess on activation patterns in verbal working memory," scientific poster at the Radiological Society of North America annual meeting, November 2005.

There are naysayers, however: Juliano and Griffiths, "A critical review of caffeine withdrawal."

2. MAKING SENSE

17   *Now smell is regarded:* Rainer W. Friedrich, "Odorant receptors make scents," *Nature* 430, 511–12 (2004).

Our thresholds for detection of many odors: J. A. Gottfried, "Smell: central nervous processing," in T. Hummel and A. Welge-Lüssen, eds., *Taste and Smell: An Update* (*Advances in Otorhinolaryngology*) (Basel, Switzerland: Karger, 2006), 44–69; Jay Gottfried, personal communication, September 2006.

Millions of olfactory nerve endings: Information about the olfactory system is from Z. Zou et al., "Odor maps in the olfactory cortex," *Proceedings of the National Academy of Sciences* 102:21, 7724–29 (2005); Z. Zou and L. B. Buck, "Combinatorial effects of odorant mixes in olfactory cortex," *Science* 311, 1477–81 (2006); R. Ranganathan and L. B. Buck, "Olfactory axon pathfinding: who is the pied piper?," *Neuron* 35:4, 599–600 (2002).

The character of a smell: A. K. Anderson et al., "Dissociated neural representations of intensity and valence in human olfaction," *Nature Neuroscience* 6:2, 196–202 (2003); Stephan Hamann, "Nosing in on the emotional brain," *Nature Neuroscience* 6, 106–8 (2003).

Its strength (how pungent?): T. W. Buchanan et al., "A specific role for the human amygdala in olfactory memory," *Learning and Memory* 10:5, 319–25 (2003); Jay Gottfried, personal communication, September 2006.

A French study in 2005: J. Plailly, "Involvement of right piriform cortex in olfactory familiarity judgments," *Neuroimage* 24, 1032–41 (2005).

18 *As one researcher said:* Tim Jacob of Cardiff University; see www.cf.ac. uk/biosi/staff/jacob/teaching/sensory/taste.html and www.cardiff.ac.uk/ biosi/staff/jacob/index.html.

Scientists have found that olfactory stimuli: S. Chu and J. J. Downes, "Odour-evoked autobiographical memories: psychological investigations of the Proustian phenomena," *Chemical Senses* 25, 111–16 (2000).

And they fall away less rapidly: C. Miles and R. Jenkins, "Recency and suffix effects with serial recall of odours," *Memory* 8:3, 195–206 (2000).

Smell memories endure: Zou et al., "Odor maps in the olfactory cortex"; Ranganathan and Buck, "Olfactory axon pathfinding"; M. Pines, "The memory of smells," in *Seeing, Hearing, and Smelling the World: A Report from the Howard Hughes Medical Institute,* at www.hhmi.org/senses/ d140.html, retrieved March 25, 2005.

Small and her colleagues . . . discovered: D. M. Small et al., "Differential neural responses evoked by orthonasal versus retronasal odorant perception in humans," *Neuron* 47, 593–605 (2005).

19 *"A key fact about":* G. M. Shepherd, "Smell images and the flavour system in the human brain," *Nature* 406, 316–21 (2006).

Though some taste cells: D. V. Smith and R. F. Margolskee, "Making sense of taste," *Scientific American,* March 2001, 32–39.

Each bud possesses up to one hundred: Bernd Lindemann, "Receptors and transduction in taste," *Nature* 413, 219–25 (2001).

20 *Even temperature enters the picture:* A. Cruz and B. G. Green, "Thermal stimulation of taste," *Nature* 403, 889–92 (2000).

In 2005, a team of researchers: K. Talavera et al., "Heat activation of TRPM5 underlies thermal sensitivity of sweet taste," *Nature* 438, 1022–25 (2005).

as Emerson described it: Ralph Waldo Emerson, *Essays and English Traits,* vol. 5, ch. 2, "Voyage to England" (Harvard Classics, 1909–14), www.bar tleby.com/5/202.html.

"Virtually every plant": Paul Breslin, personal communication, October 2006.

Scientists recently pinpointed: U.-K. Kim et al., "Genetics of human taste perception," *Journal of Dental Research* 83:6, 448–53 (2004); B. Bufe et al., "The molecular basis of individual differences in phenylthiocarbamide and propylthiouracil bitterness perception," *Current Biology* 15:4, 322–27 (2005); A. Caicedo and S. D. Roper, "Taste receptor cells that discriminate between bitter stimuli," *Science* 291, 1557–60 (2001).

21 *Breslin has found:* M. A. Sandell and P.A.S. Breslin, "Variability in a taste receptor gene determines whether we taste toxins in food"; Paul Breslin, personal communication, September 6, 2006.

22 *Our early mammalian ancestors saw:* N. J. Dominy and P. Lucas, "The ecological importance of trichromatic colour vision in primates," *Nature* 410, 363–66 (2001).

This enhanced color vision: L. A. Isbell, "Snakes as agents of evolutionary change in primate brains," *Journal of Human Evolution* 51, 1–35 (2006).

New research hints at individual variations: B. C. Verrelli and S. A. Tishkoff, "Signatures of selection and gene conversion associated with human color vision variation," *American Journal of Human Genetics* 75, 363–75 (2004).

Some percentage of women may experience: K. Jameson et al., "Richer color experience in observers with multiple photopigment opsin genes," *Psychonomic Bulletin and Review* 8:2, 244–61 (2001).

As the great psychologist William James said: William James, *Principles of Psychology,* vol. 1 (1890), http://psychclassics.yorku.ca/james/principles/prin9.htm.

23 *When we hear someone call our name:* R.A.A. Campbell and A. J. King, "Auditory neuroscience: a time for coincidence?" *Current Biology* 14, R886–88 (2004).

"Incredibly, we can detect ITDs": G. D. Pollak, "Model hearing," *Nature* 417, 502–3 (2002).

The cochlea is no passive spiral cavity: The following discussion of the auditory system derives from a personal communication with A. James Hudspeth, January 31, 2005; D. K. Chan and A. J. Hudspeth, "Ca^{2+} current-driven nonlinear amplification by the mammalian cochlea in vitro," *Nature Neuroscience* 8, 149–55 (2005); and C. Kros, "Aid from hair force," *Nature* 433, 810–11 (2005).

24 *In 2005, scientists scanned the brains:* David J. M. Kraemer et al., "Sound of silence activates auditory cortex," *Nature* 434, 158–59 (2005).

25 *When a French scientist, Gil Morrot:* G. Morrot et al., "The color of odors," *Brain and Language* 79:2, 309–20 (2001).

In one study, researchers placed monkeys: J. M. Groh et al., "Eye position influences auditory responses in primate inferior colliculus," *Neuron* 29, 509–18 (2001).

Similarly, scientists have found: Emiliano Macaluso, "Modulation of human visual cortex by crossmodal spatial attention," *Science* 289, 1206–8 (2000).

26 *Jay Gottfried and his colleagues:* J. A. Gottfried et al., "Remembrance of odors past: human olfactory cortex in crossmodal recognition memory," *Neuron* 42, 687–95 (2004).

3. WIT

27 *Though your senses are taking in:* T. Norretranders, *The User Illusion* (New York: Viking, 1998), cited in Timothy Wilson, "The adaptive unconscious: knowing how we feel," talk delivered at the Medical Center Hour, University of Virginia School of Medicine, January 21, 2004.

"We actually only see those aspects": J. Kevin O'Regan, *Research Interests,* November 2003, at http://nivea.psycho.univ-paris5.fr/TopPage/ResearchInterests.html, retrieved July 5, 2005; see also S. Yantis, "To see is to attend," *Science* 299, 54–55 (2003).

28 *This is the phenomenon demonstrated:* S. Clifasefi et al., "The effects of

alcohol on inattentional blindness," *Journal of Applied Cognitive Psychology*, DOI: 10.1002/acp.12222 (2006).

28 *Francis Crick and Christof Koch suggest:* F. Crick and C. Koch, "A framework for consciousness," *Nature Neuroscience* 6, 119–26 (2003).

Imagine this challenge: C. Sergent et al., "Timing of the brain events underlying access to consciousness during the attentional blink," *Nature Neuroscience* 8:10, 1391–99 (2005); René Marois, "Two-timing attention," *Nature Neuroscience* 8:10, 1285–86 (2005).

29 *The answer depends . . . on your interval timer:* Information on the interval timer is taken from R. B. Ivry and R.M.C. Spencer, "The neural representation of time," *Current Opinion in Neurobiology* 14, 225–32 (2004); personal communication with Richard Ivry, October 2006.

interval timing has no dedicated sensors: Ivry and Spencer, "The neural representation of time," 225; personal communication with Richard Ivry, October 2006.

the brain may judge intervals: Catalin V. Buhusi and Warren H. Meck, "What makes us tick? Functional and neural mechanisms of interval timing," *Nature Reviews Neuroscience* 6, 755–65 (2005); V. Pouthas and S. Perbal, "Time perception depends on accurate clock mechanisms as well as unimpaired attention and memory processes," *Acta Neurobiologiae Experimentalis* 64, 367–85 (2004); Uma R. Karmarka and Dean V. Buonomano, "Temporal specificity of perceptual learning in an auditory discrimination task," *Learning and Memory* 10, 141–47 (2003).

30 *Temperature may toy with this clock:* H. Woodrow, "Time perception," in S. S. Stevens, ed., *Handbook of Experimental Psychology* (New York: John Wiley, 1951), 1224–36.

When participants in one study were asked: N. Marmaras et al., "Factors affecting accuracy of producing time intervals," *Perceptual and Motor Skills* 80, 1043–56 (1995).

33 *To quantify just how efficiently:* J. Rubinstein et al., "Executive control of cognitive processes in task switching," *Journal of Experimental Psychology: Human Perception and Performance* 27:4, 763–97 (2001); see also M. A. Just et al., "Interdependence of non-overlapping cortical systems in dual cognitive tasks," *NeuroImage* 14, 417–26 (2001).

study by the National Highway Traffic Safety Administration: "Breakthrough research on real-world driver behavior released," NHTSA press release, April 20, 2006.

35 *Imaging studies by the Shaywitzes and others:* B. A. Shaywitz et al., "Disruption of posterior brain systems for reading in children with developmental dyslexia," *Biological Psychiatry* 52, 101–10 (2002); P. E. Turkeltaub, "Development of neural mechanisms for reading," *Nature Neuroscience* 6, 767–73 (2003); P. G. Simos et al., "Dyslexia-specific brain activation profile becomes normal following successful remedial training," *Neurology* 58, 1203–13 (2002).

36 *Studies show that alertness and memory:* L. Hasher et al., "It's about

time: circadian rhythms, memory, and aging," in C. Izawa and N. Ohta, eds., *Human Learning and Memory: Advances in Theory and Application* (Mahwah, N.J.: Lawrence Erlbaum Associates, 2005), 199–217.

Most of us are sharpest: Mary Carskadon, "The rhythm of human sleep and wakefulness," paper presented at the Society for Research on Biological Rhythms annual meeting, 2002.

For early risers, then, concentration tends to peak: Russell Foster and Leon Kreitzman, *Rhythms of Life* (London: Profile Books, 2004), 11.

Mary Carskadon, a chronobiologist: Carskadon, "The rhythm of human sleep and wakefulness"; M. Carskadon et al., "Adolescent sleep patterns, circadian timing, and sleepiness at a transition to early school days," *Sleep* 21:8, 871–81 (1998); M. Carskadon, ed., *Adolescent Sleep Patterns* (New York: Cambridge University Press, 2002).

How well you do at a given mental task: H.P.A. Van Dongen and D. F. Dinges, "Circadian rhythms in fatigue, alertness, and performance," in M. H. Kryger et al., *Principles and Practice of Sleep Medicine*, 3rd ed. (Philadelphia: W. B. Saunders, 2000), 391–99.

"Time-of-day effects are intriguing": Tim Salthouse, personal communication, January 28, 2005.

Scientists at the University of Pittsburgh tested: T. H. Monk et al., "Circadian rhythms in human performance and mood under constant conditions," *Journal of Sleep Research* 6:1, 9–18 (1997).

37 *On the flip side, researchers at Harvard:* K. P. Wright et al., "Relationship between alertness, performance, and body temperature in humans," *American Journal of Physiology: Regulatory, Integrative, and Comparative Physiology* 283, R1370–77 (2002).

Two mental functions may be particularly susceptible: Hasher et al., "It's about time"; L. Hasher et al., "Inhibitory control, circadian arousal, and age," in D. Gopher and A. Koriat, eds., *Attention and Performance, XVII: Cognitive Regulation of Performance: Interaction of Theory and Application* (Cambridge, Mass.: MIT Press, 1999), 653–75.

Because inhibition is particularly difficult: C. P. May, "Synchrony effects in cognition: the costs and a benefit," *Psychonomic Bulletin and Review* 6:1, 142–47 (1999).

Memory, too, may fluctuate: S. Folkard and T. H. Monk, "Time of day effects in immediate and delayed memory," in M. M. Gruneberg et al., eds., *Practical Aspects of Memory* (London: Academic Press, 1988), 142–68.

older adults tend to experience: Hasher et al., "It's about time."

38 *"It may not be a beautiful animal":* The following discussion on Kandel's life and work comes from Eric Kandel, "The molecular biology of memory storage: a dialogue between genes and synapses," *Science* 294, 1030–38 (2001); Kandel, personal communication, January 24, 2005; Eric Kandel, "Toward a biology of memory," presentation at the University of Virginia, January 28, 2005.

39 *researchers at the University of Houston:* Lisa C. Lyons et al., "Circadian modulation of complex learning in diurnal and nocturnal Aplysia," *Proceedings of the National Academy of Sciences* 102, 12589–94 (2005); see also R. I. Fernandez et al., "Circadian modulation of long-term sensitization in Aplysia," *Proceedings of the National Academy of Sciences* 100, 14415–20 (2003).

4. THE TEETH OF NOON

43 *Alessandro Benedetti asserted:* Quoted in "History of the Stomach and Intestines," www.stanford.edu/class/history13/earlysciencelab/body/stom achpages/stomachcolonintestines.html.

Not long ago, two Swiss researchers: Marianne Regard and Theodor Landis, "Gourmand syndrome: eating passion associated with right anterior lesions," *Neurology* 48, 1185–90 (1997).

45 *A recent neuroimaging study found:* A. Del Parigi et al., "Sex differences in the human brain's response to hunger and satiation," *American Journal of Clinical Nutrition* 75:6, 1017–22 (2002).

When endocrinologists at Harvard: Michael K. Badman and Jeffrey S. Flier, "The gut and energy balance: visceral allies in the obesity wars," *Science* 307, 1901–14 (2005); see also Stephen C. Woods, "Gastrointestinal Satiety Signals I: an overview of gastrointestinal signals that influence food intake," *American Journal of Physiology: Gastrointestinal and Liver Physiology* 286, G7–13 (2004).

46 *Two brain regions read this soup:* The following is from a personal communication with David Cummings, August 14, 2006.

Researchers lately confirmed that the arcuate nucleus: Roger D. Cone, "Anatomy and regulation of the central melanocortin system," *Nature Neuroscience* 8:5, 571–78 (2005).

One of the star players: M. Nakazato et al., "A role for ghrelin in the central regulation of feeding," *Nature* 409, 194–98 (2001); D. E. Cummings et al., "A preprandial rise in plasma ghrelin levels suggests a role in meal initiation in humans," *Diabetes* 50, 1714–19 (2001).

Volunteers injected with ghrelin: Y. Date et al., "The role of the gastric afferent vagal nerve in ghrelin-induced feeding and growth hormone secretion in rats," *Gastroenterology* 123:4, 1120–28 (2002).

David Cummings and his colleagues . . . see ghrelin: D. E. Cummings et al., "Ghrelin and energy balance: focus on current controversies," *Current Drug Targets* 6:2, 153–69 (2005).

46 *When the researchers measured:* D. E. Cummings et al., "A preprandial rise in plasma ghrelin levels."

47 *"The empty stomach, however":* This and all following quotes are from a personal communication with David Cummings, August 14, 2006.

Certain hormones oppose the actions: Information on leptin comes from

Heike Munzberg and Martin G. Myers Jr., "Molecular and anatomical determinants of central leptin resistance," *Nature Neuroscience* 8:5, 566–70 (2005); Michael K. Badman and Jeffrey S. Flier, "The gut and energy balance."

intake exceeds expenditure: M. Bajzer and R. J. Seeley, "Obesity and gut flora," *Nature* 444, 1009 (2006).

Leptin has worked as a therapy: Personal communication with Jeffrey Flier, July 20, 2006.

during neonatal development: J. K. Elmquist and J. S. Flier, "The fat-brain axis enters a new dimension," *Science* 304, 63–64 (2004); R. B. Simerly et al., "Trophic action of leptin on hypothalamic neurons that regulate feeding," *Science* 304, 108–10 (2004); Shirly Pinto et al., "Rapid rewiring of arcuate nucleus feeding circuits by leptin," *Science* 304, 110–15 (2004); personal communication with Jeffrey Flier, July 20, 2006.

49 *In 2005, William Carlezon and a team:* William Carlezon et al., "Antidepressant-like effects of uridine and omega-3 fatty acids are potentiated by combined treatment in rats," *Biological Psychiatry* 54:4, 343–50 (2005).

A likely explanation for this effect: This explanation and the following quotes are from a personal communication with William Carlezon, October 2006.

50 *Carlezon's finding bolsters earlier research:* Joseph R. Hibbeln, "Fish consumption and major depression," *Lancet* 351, 1213 (1998).

"This work provides more evidence": "Food ingredients may be as effective as antidepressants," press release, McLean Hospital, Harvard Medical School, February 10, 2005.

Another study suggests: S. A. Zmarzty et al., "The influence of food on pain perception in healthy human volunteers," *Physiology and Behaviour* 62:1, 185–91 (1997).

eating chocolate may create a positive mood: K. Räikkönen et al., "Sweet babies: chocolate consumption during pregnancy and infant temperament at six months," *Early Human Development* 76, 139–45 (2004).

51 *Inside the tooth and in its sockets:* The following information on teeth is from Peter W. Lucas, *Dental Functional Morphology: How Teeth Work* (New York: Cambridge University Press, 2004), 4.

By animal standards, human teeth: Peter W. Lucas, "The origins of the modern human diet," paper presented at the American Association for the Advancement of Science, February 19, 2005.

When Lieberman fed a soft diet: Interview with Dan Lieberman, February 26, 2005; D. E. Lieberman et al., "Effects of food processing on masticatory strain and craniofacial growth in a retrognathic face," *Journal of Human Evolution* 46, 655–77 (2004).

53 *Two hormones, CCK and PYY:* Stephen R. Bloom et al., "Inhibition of food intake in obese subjects by peptide YY_{3-36}," *New England Journal of Medicine* 349, 941–48 (2003).

53   *Give people an infusion:* Badman and Flier, "The gut and energy balance."

people injected with PYY: Bloom et al., "Inhibition of food intake in obese subjects."

Those rich in fiber, which move more slowly: R. L. Batterham et al., "Gut hormone PYY_{3-36} physiologically inhibits food intake," *Nature* 418, 650–54 (2002).

David Cummings and his team have shown: J. Overduin et al., "Role of the duodenum and macronutrient type in ghrelin regulation," *Endocrinology* 146:2, 845–50 (2005).

5. POST-LUNCH

54   *Some observers have even credited walking:* Osip Mandelstam, *The Noise of Time and Other Prose Pieces* (London: Quartet Books, 1988), quoted in Bruce Chatwin, *The Songlines* (New York: Penguin, 1987), 230.

55   *Around 4.2 feet per second:* R. McNeill Alexander, "Walking made simple," *Science* 308, 58–59 (2005).

Canadian scientists asked athletes: A. K. Gutmann et al., "Constrained optimization in human running," *Journal of Experimental Biology* 209, 622–32 (2006).

56   *Not long ago, biochemists . . . discovered:* A. J. Lipton et al., "S-nitrosothiols signal the ventilatory response to hypoxia," *Nature* 413, 171–74 (2001).

Brushing is not a simple matter: Kevin R. Foster, "Hamiltonian medicine: why the social lives of pathogens matter," *Science* 308, 1269–70 (2005); personal communication with Kevin Foster.

That the maw is neighborhood: Clifford Dobell, ed., *Antony van Leeuwenhoek and His Little Animals* (New York: Harcourt Brace, 1922), 239–40.

Only lately have we learned: Paul B. Eckburg et al., "Diversity of the human intestinal microbial flora," *Science* 308, 1635–38 (2005).

The . . . oral occupants are not: S. S. Socransky and A. D. Haffajee, "Dental biofilms: difficult therapeutic targets," *Periodontology* 28, 12–55 (2002).

57   *Brushing disrupts these social relationships:* Foster, "Hamiltonian medicine."

bad breath is mainly the result: Mel Rosenberg, "The science of bad breath," *Scientific American,* April 2002, 72–79; personal communication with Mel Rosenberg, July 28, 2006; "The sweet smell of Mel's success," www.taucac.org/site/News2?JServSessionIdr006=xwwa961jq1.app5b&abbr=record, retrieved July 29, 2006.

58   *The clandestine events of digestion:* William Beaumont, *Experiments and Observations on the Gastric Juice and the Physiology of Digestion* (New York: Dover Publications, 1959, reprint of 1833 edition).

a talent it owes to its inner walls: Mark Dunleavy, "Gut feeling," www.newscientist.com/lastword.

59   *The production of these juices:* M. Bouchouca et al., "Day-night patterns of gastroesophageal reflux," *Chronobiology International* 12, 267–77 (1995). *That we can digest our meals without taxing:* Quotes and explanations in the following text are from Michael Gershon, *The Second Brain* (New York: HarperCollins, 1998); M. Gershon, "The enteric nervous system: a second brain," in *Hospital Practice,* www.hosppract.com/issues/1999/07/gershon.htm.

60   *Your resident bacteria play:* The following discussion of intestinal microbes comes from: F. Bäckhed et al., "Host-bacterial mutualism in the human intestine," *Science* 307, 1915–19 (2005); L. V. Hooper and J. I. Gordon, "Commensal host-bacterial relationships in the gut," *Science* 292, 1115–18 (2001); D. R. Relman, "The human body as microbial observatory," *Nature Genetics* 30, 131–33 (2002); J.-P. Kraehenbuhl and M. Corbett, "Keeping the gut microflora at bay," *Science* 303, 1624–25 (2004); Edward Ruby et al., "We get by with a little help from our (little) friends," *Science* 303, 1305–7 (2004); L. V. Hooper et al., "Molecular analysis of commensal host-microbial relationships in the intestine," *Science* 291, 881–84 (2001); and from personal communications with Jeffrey Gordon, February 20, 2005.
      In 2005, scientists for the first time: The following description of microbial flora is from P. B. Eckburg et al., "Diversity of the human intestinal microbial flora," *Science* 308, 1635–38 (2005).

62   *Scientists at Yale discovered:* Ruslan Medzhitov, "Recognition of commensal microflora by toll-like receptors is required for intestinal homeostasis," *Cell* 118:6, 671–74 (2004); "Good bacteria trigger proteins to protect the gut," www.hhmi.org/news/medzhitov.html.

63   *germ-free mice can eat:* B. S. Samuel and J. L. Gordon, "A humanized gnotobiotic mouse model of host-archaeal-bacterial mutualism," *Proceedings of the National Academy of Sciences* 103:26, 10011–16 (2006); F. Bäckhed et al., "The gut microbiota as an environmental factor that regulates fat storage," *Proceedings of the National Academy of Sciences* 101:44, 15718–23 (2004).
      Gordon and his lab mates: R. E. Ley et al., "Human gut microbes associated with obesity," *Nature* 444, 1022–23 (2006); P. J. Turnbaugh et al., "An obesity-associated gut microbiome with increased capacity for energy harvest," *Nature* 444, 1027–31 (2006).

64   *How long it takes for bowel, bugs, and brain:* R. H. Goo et al., "Circadian variation in gastric emptying of meals in man," *Gastroenterology* 93, 513–18 (1987).
      Franz Halberg . . . determined: Franz Halberg et al., "Chronomics: circadian and circaseptan timing of radiotherapy, drugs, calories, perhaps nutriceuticals and beyond," *Journal of Experimental Therapeutics and Oncology* 3:5, 223 (2003).
      some of those peripheral clocks: Karl-Arne Stokkan et al., "Entrainment of the circadian clock in the liver by feeding," *Science* 291, 490–93 (2001).

64 *One recent study showed:* Ueli Schibler et al., "Peripheral circadian oscillators in mammals: time and food," *Journal of Biological Rhythms* 18:3, 250–60 (2003); J. Rutter et al., "Regulation of clock and NPAS2 DNA binding by the redox state of NAD cofactors," *Science* 293, 510–14 (2001).

65 *gastroenterologists hurdled the obstacles:* C.S.J. Probert et al., "Some determinants of whole-gut transit time: a population-based study," *Quarterly Journal of Medicine* 88, 311–15 (1995).
"A meal is typically a mixture": R. Bowen, "Gastrointestinal transit: how long does it take?," www.vivo.colostate.edu/hbooks/pathphys/digestion/basics/transit.html, retrieved September 29, 2006; personal communication with Richard Bowen, October 2006.

66 *Feces . . . are made mostly of:* Ralph A. Lewin, *Merde: Excursions in Scientific, Cultural, and Socio-Historical Coprology* (New York: Random House, 1999); Bäckhed, "Host-bacterial mutualism in the human intestine," 1917.
The stink in feces: Bill Rathmell, "No Bull," www.newscientist.com.
the compound is said to be used: K. G. Friedeck, "Soy protein fortification of a low-fat dairy-based ice cream," *Journal of Food Science* 68, 2651 (2003).
Scientists investigated the phenomenon: Michael D. Levitt et al., "Evaluation of an extremely flatulent patient," *American Journal of Gastroenterology* 93:11, 2276–81 (1998).

67 *Just by living, by keeping your heart beating:* The following discussion of metabolism comes from Eric Ravussin, "A NEAT way to control weight?," *Science* 307, 530–31 (2005); personal communication with Eric Ravussin, August 8, 2006.

68 *unless they're pregnant or nursing:* Jean Mayer, *Human Nutrition* (Springfield, Ill.: Charles C. Thomas, 1979), 21–24.
Scientists at Harvard have shown: Eric S. Bachman, "ßAR signaling required for diet-induced thermogenesis and obesity resistance," *Science* 297, 843–45 (2002).
One of the genes responsible: Bradford B. Lowell and Bruce M. Spiegelman, "Towards a molecular understanding of adaptive thermogenesis," *Nature* 404, 652–60 (2000).

69 *In one two-month study, scientists at the Mayo Clinic:* J. A. Levine et al., "Role of nonexercise activity thermogenesis in resistance to fat gain in humans," *Science* 283, 212–14 (1999); James Levine and Michael Jensen, response to "A fidgeter's calculation," *Science* 284, 1123 (2000).
the Mayo Clinic team set out to pinpoint: J. A. Levine et al., "Interindividual variation in posture allocation: possible role in human obesity," *Science* 307, 584–86 (2005).

6. THE DOLDRUMS

73 *It's the doldrums:* Norton Juster, *The Phantom Tollbooth* (New York: Random House/Bullseye Books, 1988), 24.

74 *This and other questions of weariness:* Eighth annual meeting of the Society for Research on Biological Rhythms, Amelia Island, Florida, 2002 (hereafter, SRBR meeting, 2002).

Soon to speak was Mary Carskadon: Mary Carskadon, "Guidelines for the Multiple Sleep Latency Test (MSLT): a standard measure of sleepiness," *Sleep* 9, 519–24 (1986); Mary Carskadon and William Dement, "Daytime sleepiness: quantification of a behavioral state," *Neuroscience Biobehavioral Review* 11, 307–17 (1987).

On the 7-point Stanford Sleepiness Scale: E. Hoddes et al., "Qualification of sleepiness: a new approach," *Psychophysiology* 10, 431–36 (1973).

According to neuroscientists, a yawn: A. Argiolas and M. R. Melis, "The neuropharmacology of yawning," *European Journal of Pharmacology* 343:1, 1–16 (1998).

75 *But when Robert Provine . . . tested this theory:* R. Provine, "Yawning: no effect of 3–5% CO_2, 100% O_2, and exercise," *Behavioral Neural Biology* 48:3, 382–93 (1987).

As Dr. Seuss said: Dr. Seuss's Sleep Book (New York: Random House, 1962).

To probe the nature of contagious yawning: S. M. Platek et al., "Contagious yawning: the role of self-awareness and mental state attribution," *Cognitive Brain Research* 17, 223–27 (2003).

A follow-up fMRI study: S. Platek et al., "Contagious yawning and the brain," *Cognitive Brain Research* 23, 448–52 (2005); personal communication with Platek, September 7, 2006.

76 *A deep seasonal rhythm largely ignored:* N. E. Rosenthal, *Winter Blues: Seasonal Affective Disorder* (New York: Guilford Press, 1998), 287 f.

78 *"Gastric stretch" is thought to have:* S. Schacter et al., "Vagus nerve stimulation," *Epilepsia* 39, 677–86 (1998); A. Yamanaka et al., "Hypothalamic orexin neurons regulate arousal according to energy balance in mice," *Neuron* 38, 701–13 (2003).

In cats, the mere act of gently stimulating: T. Kukorelli and G. Juhasz, "Sleep induced by intestinal stimulation in cats," *Physiology and Behavior* 19, 355–58 (1977).

A big meal rich in fat: A. Wells et al., "Influence of fat and carbohydrate on postprandial sleepiness, mood, and hormones," *Physiology and Behavior* 61:5, 679–86 (1997).

When scientists compared midafternoon: Gary Zammit et al., "Postprandial sleep in healthy men," *Sleep* 18:4, 229–31 (1995).

The work of Carskadon and others: M. A. Carskadon and C. Acebo, "Regulation of sleepiness in adolescents: update, insights, and speculation," *Sleep* 25:6, 606–14 (2002); M. Carskadon, "The rhythm of human sleep and wakefulness," presentation at SRBR meeting, 2002; W. Dement and C. Vaughan, *The Promise of Sleep* (New York: Dell, 2000), 79–84.

Peretz Lavie confirmed this: Peretz Lavie, *The Enchanted World of Sleep* (New Haven: Yale University Press, 1996), 51; personal communication with Lavie, February 14, 2005.

79   *Dale Edgar . . . verified the location:* D. M. Edgar et al., "Effect of SCN le-
      sions on sleep in squirrel monkeys: evidence for opponent processes in
      sleep-wake regulation," *Journal of Neuroscience* 13, 1065–79 (1993); De-
      ment and Vaughan, *The Promise of Sleep,* 78–81. personal communica-
      tion with Dement, March 5, 2005.
      Just how severely you suffer: M. Carskadon, "The rhythm of human sleep
      and wakefulness."
80   *Studies of fatigue-related accidents:* M. M. Mitler et al., "Catastrophes,
      sleep, and public policy: consensus report," *Sleep* 11, 100–109 (1988).
      At around 4 P.M., drivers are: Jim Horne and Louise Reyner, "Vehicle ac-
      cidents related to sleep: a review," *Occupational and Environmental Med-
      icine* 56, 289–94 (1999).
81   *Napping is common in traditional cultures:* Wilse B. Webb and David
      F. Dinges, "Cultural perspectives on napping and the siesta," in David
      Dinges, ed., *Sleep and Alertness* (New York: Raven Press, 1989), 247–65.
      "You must sleep sometime": Churchill quoted at www.powerofsleep.org/
      sleepfacts.htm and at www.mysleepcenter.com/sleepquotations.html.
      Claudio Stampi, an Italian sleep researcher: Claudio Stampi, *Why We Nap*
      (Boston: Birkhauser, 1992).
      As . . . William Dement points out: Dement and Vaughan, *The Promise of
      Sleep,* 371–77.
82   *researchers at NASA tested:* Dement and Vaughan, *The Promise of Sleep,*
      374; see also, M. R. Rosekind et al., "Crew factors in flight operations
      IX: effects of planned cockpit rest on crew performance and alertness
      in long-haul operations," NASA Technical Memorandum 108839 (Moffett
      Field, Calif.: NASA Ames Research Center, 1994).
      "Everyone knows about the need": F. Turek, "Future directions in circa-
      dian and sleep research," presentation at SRBR meeting, 2002.
      Even for those of us with lives: M. Takahashi et al., "Maintenance of alert-
      ness and performance by a brief nap after lunch under prior sleep defi-
      cit," *Sleep* 23:6, 813–19 (2000); S.M.W. Rajaratnam and J. Arendt, "Health
      in a 24-h society," *Lancet* 358, 999–1005 (2001).
      for sleepy subjects taking monotonous: J. A. Horne and L. A. Reyner,
      "Counteracting driver sleepiness: effects of napping, caffeine, and pla-
      cebo," *Psychophysiology* 33:3, 306–9 (1996).
      Japanese researchers who recently conducted: M. Takahashi et al., "Post-
      lunch nap as a worksite intervention to promote alertness on the job,"
      Ergonomics 47:9, 1003–13 (2004).
      Sara Mednick and her colleagues: S. C. Mednick et al., "The restorative
      effect of naps on perceptual deterioration," *Nature Neuroscience* 5, 677–
      81 (2002); P. Maquet, "Be caught napping: you're doing more than rest-
      ing your eyes," *Nature Neuroscience* 5, 618–19 (2002).
83   *Mednick showed that naps:* S. Mednick et al., "Sleep-dependent learning:
      a nap is as good as a night," *Nature Neuroscience* 6, 697–98 (2003).
      In early 2007 came news: A Naska et al., "Siesta in healthy adults and cor-

onary mortality in the general population," *Archives of Internal Medicine* 167, 296–301 (2007).

In short, say sleep researchers: Personal communication with Sara Mednick, October 3, 2006; Dement and Vaughan, *The Promise of Sleep,* 371.

The latest siesta studies suggest: A. Brooks and L. Lack, "A brief afternoon nap following nocturnal sleep restriction: which nap duration is most recuperative?," *Sleep* 29:6, 831–40 (2006).

The human body is "programmed" for a siesta: M. Carskadon, "Ontogeny of human sleepiness as measured by sleep latency," in D. F. Dinges and R. J. Broughton, eds., *Sleep and Alertness: Chronobiological, Behavioral, and Medical Aspects of Napping* (New York: Raven Press, 1989), 53–69.

7. STRUNG OUT

84 *William James once wrote:* William James, *The Principles of Psychology,* vol. 2, 1890, 415–16; http://psychclassics.yorku.ca/james/principles/prin25.htm.

86 *LeDoux has teased out:* The following description of fear and the brain comes from "Neurosystems underlying fear," paper delivered at the symposium "Stress and the Brain," National Institutes of Health, Washington, D.C., March 12, 2003; E. K. Lanuza et al., "Unconditioned stimulus pathways to the amygdala: effects of posterior thalamic and cortical lesions on fear conditioning," *Neuroscience* 125, 305–15 (2004); J. LeDoux, "The emotional brain, fear, and the amygdala," *Cellular and Molecular Neurobiology* 23:4–5, 727–38 (2003); and personal communication with Joseph LeDoux, January 16, 2005.

87 *The "look out!" message:* The following description of the fight-flight response is from Bruce McEwen, *The End of Stress* (Washington, D.C.: Dana Press, 2002).

"The idea of this activity": The following description and quotes throughout this chapter derive from McEwen, *The End of Stress,* and personal communication with Bruce McEwen, January 17, 2005.

88 *A Hungarian scientist . . . was the first:* H. Selye, "A syndrome produced by diverse nocuous agents," *Nature* 138, 32 (1936).

89 *These days, scientists tend to define:* Robert Sapolsky, "Sick of poverty," *Scientific American,* December 2005, 96.

90 *McEwen and his colleague:* E. S. Epel et al., "Stress and body shape: stress-induced cortisol secretion is consistently greater among women with central fat," *Psychosomatic Medicine* 62:5, 623–32 (2000).

91 *That changed with research from UCSF:* Mary F. Dallman, "Chronic stress and obesity: a new view of comfort food," *Proceedings of the National Academy of Sciences* 100:20, 11696–11701 (2003); Norman Pecoraro et al., "Chronic stress promotes palatable feeding, which reduces signs of stress: feedforward and feedback effects of chronic stress," *Endocrinology* 145, 3754 (2004); Mary Dallman, "Glucocorticoids: food intake, abdomi-

nal obesity and wealthy nations in 2004," *Endocrinology* 145, 2633 (2004).
91 *When scientists . . . studied rats:* K. Kamara et al., "High-fat diets and
stress responsivity," *Physiology and Behavior* 64, 1–6 (1998).
Some 150 studies suggest: D. A. Padgett and R. Glaser, "How stress influences the immune response," *Trends in Immunology* 24:8, 444–48 (2003).
People who endure stressful conditions: S. Cohen et al., "Psychological
stress and susceptibility to the common cold," *New England Journal of
Medicine* 325, 606–12 (1991).
They're also apt to produce: When Ronald Glaser and Janice Kiecolt-
Glaser and their team of researchers at Ohio State University studied
the effect of stress on the body's ability to respond to the challenge of
a vaccine, they found that stressed-out medical students produced a
weak antibody response to the hepatitis B vaccine compared with control subjects, and caretakers of people with Alzheimer's disease showed
a dampened response to a flu virus vaccine. J. K. Kiecolt-Glaser et al.,
"Stress-induced modulation of the immune response to recombinant
hepatitis B vaccine," *Psychosomatic Medicine* 54, 22–29 (1992); "Chronic
stress alters the immune response to influenza virus vaccine in older
adults," *Proceedings of the National Academy of Sciences* 93, 3043–47
(1996).
One study found that in women caring for a relative: J. K. Kiecolt-Glaser et
al., "Slowing of wound healing by psychological stress," *Lancet* 346, 1194–
96 (1995). Glaser's team also found that minor wounds inflicted on the
hard palate of dental students three days before a major test healed an
average of 40 percent slower than wounds in the same subjects during
summer vacation. P. T. Marucha et al., "Mucosal wound healing is impaired by examination stress," *Psychosomatic Medicine* 60, 362–65 (1998).
Psychological stress . . . inhibits a key component: R. Glaser et al., "Stress-
related changes in proinflammatory cytokine production in wounds,"
Archives of General Psychiatry 56, 450–56 (1999).
92 *even temporary stress, if sufficiently severe:* Ajai Vyas et al., "Chronic stress
induces contrasting patterns of dendritic remodeling in hippocampal
and amygdaloid neurons," *Journal of Neuroscience* 22:15, 6810–18 (2002);
personal communication with Bruce McEwen, January 17, 2005.
A kind of reverse process goes on: R. Pawlak, "Tissue plasminogen activa-
tor in the amygdala is critical for stress-induced anxiety-like behavior,"
Nature Neuroscience 6:2, 168–74 (2003).
"People who are stressed over long periods": E. S. Epel, "Accelerated telo-
mere shortening in response to life stress," *Proceedings of the National
Academy of Sciences,* DOI: 10.1073/pnas.0407162101 (2004).
94 *Whether or not we are derailed:* A. Caspi et al., "Influence of life stress on
depression: moderation in the 5-HTT gene," *Science* 301, 386–89 (2003);
Stephan Hamann, "Blue genes: wiring the brain for depression," *Nature
Neuroscience* 8:6, 701 (2005).
The genes made headlines: Peter Kramer, "Tapping the mood gene," *New
York Times,* July 26, 2003, A13.

Try to feel in control: Esther Sternberg, personal communication, January 17, 2005.

95 *Richard Davidson . . . and his colleagues:* R. Davidson et al., "Alterations in brain and immune function produced by mindfulness meditation," *Psychosomatic Medicine* 65, 564–70 (2003).

mindfulness meditation can be a powerful: J. Kabat-Zinn, "Mindfulness-based stress reduction: past, present and future," *Clinical Psychology Science and Practice* 10, 144–56 (2003); J. Kabat-Zinn, "Influence of a mindfulness-based stress reduction intervention on rates of skin clearing in patients with moderate to severe psoriasis undergoing phototherapy (UVB) and photochemotherapy (PUVA)," *Psychosomatic Medicine* 60, 625–32 (1998).

96 *Music with a quick tempo:* C. L. Krumhansl, "An exploratory study of musical emotions and psychophysiology," *Canadian Journal of Experimental Psychology* 51:4, 336–53 (1997).

scientists at the Montreal Neurological Institute: A. J. Blood and R. J. Zatorre, "Intensely pleasurable responses to music correlate with activity in brain regions implicated in reward and emotion," *Proceedings of the National Academy of Sciences* 98:20, 11818–23 (2001).

Other studies suggest that music: J. A. Etzel et al., "Cardiovascular and respiratory responses during musical mood induction," *International Journal of Psychophysiology* 61, 57–59 (2006).

Dairy cows make more milk: A. North and L. MacKenzie, "Milk yields affected by music tempo," *New Indian Express,* July 4, 2001.

97 *People with strong social:* McEwen, *The End of Stress,* 145.

Allan Reiss and his colleagues: Dean Mobbs et al., "Humor modulates the mesolimbic reward centers," *Neuron* 40, 1041–48 (2003).

That humor sparks the brain's primeval: Jaak Panksepp, "Beyond a joke: from animal laughter to human joy," *Science* 308, 62–63 (2005).

"Humor can be dissected": E. B. White, *A Subtreasury of American Humor* (New York: Coward-McCann, 1941), xvii.

What most powerfully affects: Bruce McEwen, personal communication, January 17, 2005.

8. IN MOTION

98 *More than a hundred studies:* R. K. Dishman, "Neurobiology of exercise," *Obesity* 14:3, 345–56 (2006); D. M. Landers, "The influence of exercise on mental health," *President's Council on Physical Fitness and Sports Research Digest* 2:12 (1997), www.fitness.gov/mentalhealth.htm; B. S. Hale et al., "State anxiety responses to 60 minutes of cross training," *British Journal of Sports Medicine* 36, 105–7 (2002).

researchers put young musicians: D. Wasley and A. Taylor, "The effect of physical activity and fitness on psycho-physiological responses to a musical performance and laboratory stressor," in K. Stevens et al., eds., *Pro-*

ceedings of the 7th International Conference on Music Perception and Cognition (Sydney, Australia: Casual Productions, 2002), 93–96.

99 *It's true that prolonged cardiovascular exercise:* M. T. Ruffin et al., "Exercise and secondary amenorrhoea linked through endogenous opioids," *Sports Medicine* 10:2, 65–71 (1994).

It's also true that a rise in endorphins: M. Daniel et al., "Opiate receptor blockade by naltrexone and mood state after acute physical activity," *British Journal of Sports Medicine* 26:2, 111–15 (1992).

But it remains unclear: G. A. Sforzo, "Opioids and exercise. An update," *Sports Medicine* 7:2, 109–24 (1989); John Ratey, *A User's Guide to the Brain* (New York: Vintage, 2001), 360.

The uplifting effect may be due to: Ratey, *A User's Guide to the Brain,* 360. Pretty et al., "The mental and physical outcomes of green exercise," *International Journal of Environmental Health Research* 15:5, 319–37 (2005); J. Baatile et al., "Effect of exercise on perceived quality of life of individuals with Parkinson's disease," *Journal of Rehabilitation Research and Development* 37:5, 529–34 (2000); A. A. Bove, "Increased conjugated dopamine in plasma after exercise training," *Journal of Laboratory and Clinical Medicine* 104:1, 77–85 (1984).

Most likely the mood lift comes: Ratey, *A User's Guide to the Brain,* 360.

In a study called SMILE*:* M. Bibyak et al., "Exercise treatment for major depression: maintenance of therapeutic benefit at 10 months," *Psychosomatic Medicine* 62, 633–38 (2000).

One study found that patients: Andrea L. Dunn et al., "Exercise treatment for depression: efficacy and dose response," *American Journal of Preventive Medicine* 28:1, 1–8 (2005).

A longitudinal survey of . . . 6,800 men and women: D. I. Galper et al., "Inverse association between physical inactivity and mental health in men and women," *Medicine and Science in Sports and Exercise* 38:1, 173–78 (2006).

100 *Blumenthal suspects that people who exercise:* James Blumenthal, personal communication, August 7, 2006.

Late afternoon and early evening are considered: The following description of exercise rhythms comes from "Circadian rhythms in sports performance," in T. Reilly et al., *Biological Rhythms and Exercise* (New York: Oxford University Press, 1997); personal communication with Thomas Reilly, September 2006; C. M. Winget et al., "Circadian rhythms and athletic performance," *Medicine and Science in Sports and Exercise* 17, 498–516 (1985).

Airways are most open late: Boris I. Medarov, study presented at the 70th annual international scientific assembly of the American College of Chest Physicians, October 23–28, 2004, in Seattle.

101 *back pain is often less severe:* Michael Smolensky and Lynne Lamberg, *The Body Clock Guide to Better Health* (New York: Holt, 2000), 223–26.

Even very young infants: A. N. Meltzoff, "Elements of a developmental theory of imitation," in A. N. Meltzoff and W. Prinz, eds., *The Imitative Mind: Development, Evolution, and Brain Bases* (Cambridge: Cambridge University Press, 2002), 19–41.

New research suggests that our brains: M. Iacoboni, "Understanding others: imitation, language, empathy," in S. Hurley and N. Chater, eds., *Perspectives on Imitation: From Cognitive Neuroscience to Social Science* (Cambridge, Mass.: MIT Press, in press), www.cbd.ucla.edu/bios/royaumont .pdf.

102 *Some modern researchers argue that only sustained:* www.cdc.gov/nccd php/dnpa/physical/recommendations/index.htm.

103 *Australian scientists persuaded a dozen men:* S. M. Gunn et al., "Determining energy expenditure during some household and garden tasks," *Medicine and Science in Sports and Exercise* 34:5, 895–902 (2002).

Climbing stairs counts too: K. C. The and A. R. Aziz, "Heart rate, oxygen uptake, and energy cost of ascending and descending the stairs," *Medicine and Science in Sports and Exercise* 34:4, 695–99 (2002). To maximize benefits, the scientists recommended that people climb up and down the twenty-two flights seven times at each session, for a total activity period of twenty-six minutes, four times a week.

only a quarter of all American adults: Centers for Disease Control, Morbidity and Mortality Weekly Report, December 1, 2005.

Evidence suggests that early hunter-gatherers walked: Personal communication with Richard Wrangham and Dan Lieberman, Harvard University, February 26, 2005.

Loss of muscle and bone: Miriam Nelson, personal communication, October 30, 2006.

104 *Resistance training works its muscle magic:* Henning Wackerhage, personal communication, October 2006.

The proteins that make up: G. Biolo et al., "Increased rates of muscle protein turnover and amino acid transport after resistance exercise in humans," *American Journal of Physiology, Endocrinology, and Metabolism* 268, E514–20 (2005).

A team of scientists at the Cleveland Clinic: V. K. Ranganathan et al., "From mental power to muscle power—gaining strength by using the mind," *Neuropsychologia* 42, 944–56 (2004).

105 *To keep your muscle and bone mass:* J. E. Layne and M. Nelson, "The effects of progressive resistance training on bone density. A review," *Medicine and Science in Sports and Exercise* 31:1, 25–30 (1999).

In 2005, a team from the University of Massachusetts: P. M. Clarkson et al., "Variability in muscle size and strength gain after unilateral resistance training," *Medicine and Science in Sports and Exercise* 37:6, 964–72 (2005).

Muscle soreness generally peaks: H. Wackerhage, "Recovering from eccentric exercise: get weak to become strong," *Journal of Physiology* 553, 681 (2003).

105 *Stretching does not prevent it:* R. Herbert and M. Gabriel, "Effects of stretching before and after exercising on muscle soreness and risk of injury: systematic review," *British Medical Journal* 325, 468 (2002).

106 *To teach this lesson:* Henning Wackerhage, personal communication, October 2006.

Delayed muscle soreness is caused by: J. Fridén and R. L. Lieber, "Eccentric exercise–induced injuries to contractile and cytoskeletal muscle fibre components," *Acta Physiologica Scandinavica* 171, 321–26 (2001).

On the bright side, muscles respond: P. M. Clarkson, "Molecular responses of human muscle to eccentric exercise," *Journal of Applied Physiology* 95, 2485–94 (2003); P. M. Clarkson and I. Tremblay, "Exercise-induced muscle damage, repair, and adaptation in humans," *Journal of Applied Physiology* 65:1, 1–6 (1988).

Our species is built for this: D. M. Bramble and D. E. Lieberman, "Endurance running and the evolution of *Homo*," *Nature* 432, 345–52 (2004); personal communication with Dan Lieberman, January 2005.

107 *Using a treadmill with a force plate:* P. Weyand et al., "Faster top running speeds are achieved with greater ground forces, not more rapid leg movements," *Journal of Applied Physiology* 89, 1991–2000 (2000).

British scientists showed that blood: Philip J. Kilner et al., "Asymmetric redirection of flow through the heart," *Nature* 404, 759–61 (2000).

108 *Researchers have found that the drag:* J. Y. Ji et al., "Shear stress causes nuclear localization of endothelial glucocorticoid receptor and expression from the GRE promoter," *Circulation Research* 92, 279 (2003).

In addition, even lower-intensity exercise: R. Rauramaa, "Results of DNASCO (DNA polymorphism and carotid atherosclerosis) study, a six-year study on the effects of low-intensity exercise and genetic factors on atherosclerosis" (abstract 3855), presented at the American Heart Association's Scientific Sessions Conference, 2001.

Miller used clips from the movie: M. Miller et al., "Impact of cinematic viewing on endothelial function," *Heart* 92, 261–62 (2006); personal communication with Michael Miller, September 2006.

110 *In fact, according to Timothy Noakes:* T. D. Noakes and A. St. Clair Gibson, "Logical limitations to the 'catastrophe' models of fatigue during exercise in humans," *British Journal of Sports Medicine* 38, 648–49 (2004); personal communication with Timothy Noakes, August 2006.

"No study has yet clearly established": A. St. Clair Gibson and T. D. Noakes, "Evidence for complex system integration and dynamic neural regulation of skeletal muscle recruitment during exercise in humans," *British Journal of Sports Medicine* 38, 797–806 (2004); Noakes and St. Clair Gibson, "Logical limitations to the 'catastrophe' models of fatigue."

To demonstrate the mental component of exhaustion: D. A. Baden et al., "Effect of anticipation during unknown or unexpected exercise duration on rating of perceived exertion, affect, and physiological function," *British Journal of Sports Medicine* 39, 742–46 (2005); A. St. Clair Gibson et al., "The role of information processing between the brain and peripheral physiological systems in pacing and perception of effort," *Sports Medicine* 36:8, 705–22 (2006).

111 *One possibility is a molecule called interleukin-6:* P. J. Robson-Ansley et al., "Acute interleukin-6 administration impairs athletic performance in healthy, trained male runners," *Canadian Journal of Applied Physiology* 29:4, 21–24 (2004). See also B. K. Pedersen and M. Febbraio, "Muscle-derived interleukin-6: a possible link between skeletal muscle, adipose tissue, liver, and brain," *Brain, Behavior, and Immunity* 19, 371–76 (2005).

112 *reduces the incidence of colds:* C. Ulrich et al., "Moderate-intensity exercise reduces the incidence of colds in postmenopausal women," *American Journal of Medicine* 119: 11, 937–42 (2006).

In 2004, a team of Japanese researchers: Koji Okamura et al., presentation at Experimental Biology 2004 meeting, April 17–21, 2004, Washington, D.C.

New studies show that this stepped-up: E. Borsheim and R. Bahr, "Effect of exercise intensity, duration and mode on post-exercise oxygen consumption," *Sports Medicine* 33:14, 1037–60 (2003).

The Amish people . . . beautifully demonstrate: D. Bassett et al., "Physical activity in an Old Order Amish community," *Medicine and Science in Sports and Exercise* 36:1, 79–85 (2004).

Researchers have calculated: J. O. Hill et al., "Obesity and the environment: where do we go from here?," *Science* 299, 853–55 (2003).

Some years ago, the brain researcher: H. van Praag, "Running enhances neurogenesis, learning, and long-term potentiation in mice," *Proceedings of the National Academy of Sciences* 96, 13427–31 (1999).

113 *a molecule so important in helping brain cells:* Carl Cotman interview on *The Health Report*, ABC Radio National, Monday, March 24, 1997.

"It's reasonable to speculate": Personal communication with Art Kramer, January 16, 2005.

Not long ago, Naftali Raz: Naftali Raz et al., "Regional brain changes in aging healthy adults: general trends, individual differences, and modifiers," *Cerebral Cortex* 15:11, 1676–89 (2005); personal communication with Raz, February 3, 2005.

114 *Tim Salthouse of the University of Virginia:* Tim Salthouse, personal communication, January 28, 2005.

One big Canadian study: D. Laurin et al., "Physical activity and risk of cognitive impairment and dementia in elderly persons," *Archives of Neurology* 58, 498–504 (2001).

This was confirmed in 2004: J. Weuve et al., "Physical activity, including walking, and cognitive function in older women," *Journal of the American Medical Association* 292:12, 1454–61 (2004).

Art Kramer and his colleagues recently investigated: Art Kramer, personal communication, January 16, 2005; S. Colcombe and A. F. Kramer, "Fitness effects on the cognitive function of older adults: a meta-analytic study," *Psychological Science* 14, 125–30 (2003); J. D. Churchill et al., "Exercise, experience, and the aging brain," *Neurobiology of Aging* 23, 941–55 (2002); A. F. Kramer et al., "Aging, fitness and neurocognitive function," *Nature* 400, 418–19 (1999).

9. PARTY FACE

119   *"Give me a bowl of wine"*: *Julius Caesar*, act 4, scene 3.

  Time of day influences how quickly: J. Wasielewski and F. Holloway, "Alcohol's interactions with circadian rhythms," *Alcohol Research and Health* 25:2, 94–100 (2001).

  In one study of twenty men: N. W. Lawrence et al., "Circadian variation in effects of ethanol in man," *Pharmacology, Biochemistry, and Behavior* 18 (supp. 1), 555–58 (1983); see also J. Brick et al., "Circadian variations in behavioral and biological sensitivity to ethanol," *Alcoholism: Clinical and Experimental Research* 8, 204–11 (1984).

  This time of day may in fact affect: T. Reilly et al., *Biological Rhythms and Exercise* (New York: Oxford University Press, 1997), 40–41.

120   *Drugs such as marijuana and hashish:* L. D. Chait, "Acute and residual effects of alcohol and marijuana, alone and in combination, on mood and performance," *Psychopharmacology* 115, 340–49 (1994).

  William James wrote about the "curious increase": William James, *The Principles of Psychology,* vol. 1, 639 (1890), http://psychclassics.yorku.ca/james/principles/prin15.htm.

  alcohol can either lessen stress or intensify it: M. A. Sayette, "Does drinking reduce stress?," *Alcohol Research and Health* 23:4, 250–55 (1999); M. A. Sayette, "An appraisal-disruption model of alcohol's effects on stress responses in social drinkers," *Psychological Bulletin* 114, 459–76 (1993); Michael Sayette, personal communication, August 2006.

121   *After a person starts drinking:* P. N. Friel et al., "Variability of ethanol absorption and breath concentrations during a large-scale alcohol administration study," *Alcoholism: Clinical and Experimental Research* 19:4, 1055 (1995).

  An hour after consuming: "Alcohol and transportation safety," Alcohol Alert 52, National Institute on Alcohol Abuse and Alcoholism, April 2001.

  Women reach higher peak blood alcohol levels: M. Mumenthaler et al., "Gender differences in moderate drinking effects," *Alcohol Research and Health* 23:1, 55–64 (1999).

  work by scientists at Stanford University: Mumenthaler et al., "Gender differences in moderate drinking effects," 57.

  in the sleep-deprived, alcohol hits hard: T. Roehrs et al., "Sleep extension, enhanced alertness and the sedating effects of ethanol," *Pharmacology, Biochemistry, and Behavior* 34, 321–24 (1989).

122   *William James described this failure:* James, *The Principles of Psychology,* vol. 1, 251.

It's one of the "seven sins of memory": Daniel Schacter, *The Seven Sins of Memory* (Boston: Houghton Mifflin, 2001), 63; A. Maril et al., "On the tip of the tongue: an event-related fMRI study of semantic retrieval failure and cognitive conflict," *Neuron* 31, 653–60 (2001).

Or perhaps a technological solution: "On the tip of my tongue," *New Scientist* 7, 17 (2002).

"Real world tests of automated face-recognition systems": P. Sinha, "Recognizing complex patterns," *Nature Neuroscience Supplement* 5, 1093–97 (2002).

No, wrote Milan Kundera: Milan Kundera, *Immortality* (New York: Perennial, 1999), 13.

124 *Some cases of prosopagnosia:* B. C. Duchaine and K. Nakayama, "Developmental prosopagnosia: a window to content-specific face processing," *Current Opinion in Neurobiology* 16, 166–73 (2006); Brad Duchaine, personal communication, August 2006.

Imaging studies show that neural activity: D. Y. Tsao et al., "A cortical region consisting entirely of face-selective cells," *Science* 311, 670–74 (2006); G. Loffler, "fMRI evidence for the neural representation of faces," *Nature Neuroscience* 8:10, 1386–90 (2005).

In a recent experiment with monkeys: D. Y. Tsao, "A dedicated system for processing faces," *Science* 314, 72–73 (2006).

Indeed, it appeared to be pretty far-fetched: R. Quian Quiroga et al., "Invariant visual representation by single neurons in the human brain," *Nature* 435, 1102–7 (2005).

125 *"This neuron looks for all the world":* C. E. Connor, "Friends and grandmothers," *Nature* 435, 1036–37 (2005).

"I suspect . . . that if this patient": Christof Koch, personal communication, September 2006.

Doris Tsao's work suggests: Tsao, "A dedicated system for processing faces," 72–73.

The Tierra del Fuegans have an expression: Howard Rheingold, *They Have a Word for It* (Louisville, Ky.: Sarabande Books, 2000), 80.

The whites of our eyes, which highlight: H. Kobayashi and S. Kohshima, "Unique morphology of the human eye and its adaptive meaning: comparative studies on external morphology of the primate eye," *Journal of Human Evolution* 40, 419–35 (2001).

126 *A team at University College London:* K. Kampe et al., "Reward value of attractiveness and gaze," *Nature* 413, 589 (2001).

Most of us prefer faces: L. Mealey et al., "Symmetry and perceived facial attractiveness," *Journal of Personality and Social Psychology* 76, 151–58 (1999).

A team of Scottish and Japanese scientists: D. Perrett et al., "Effects of sexual dimorphism on facial attractiveness," *Nature* 394, 884–87 (1998).

Craig Roberts and his team: S. C. Roberts et al., "Female facial attractiveness increases during the fertile phase of the menstrual cycle," *Proceedings of the Royal Society of London B* (Supp.), DOI: 10.1098/rsbl.2004.0174

(2004); I. S. Penton-Voak et al., "Menstrual cycle alters face preference," *Nature* 399, 741–42 (1999); Craig Roberts, personal communication, January 21, 2005.

127 *"In my lectures, I ask whether"*: personal communication, Mel Rosenberg, September 2006.

To determine whether attractiveness: F. Thorne et al., "Effects of putative male pheromones on female ratings of male attractiveness: influence of oral contraceptives and the menstrual cycle," *Neuroendocrinology Letters* 23:4, 291–97 (2002).

our olfactory system is exquisitely sensitive: R. W. Friedrich, "Odorant receptors make scents," *Nature* 430, 511–12 (2004).

Women are better at the task than men: P. Dalton et al., "Gender-specific induction of enhanced sensitivity to odors," *Nature Neuroscience* 5, 199–200 (2002).

128 *According to D. Michael Stoddart:* D. M. Stoddart, *The Scented Ape* (New York: Cambridge University Press, 1991); personal communication, D. M. Stoddart, March 3, 2005.

wicked out into the world: Charles Wysocki and George Preti, "Facts, fallacies, fears, and frustrations with human pheromones," *Anatomical Record* 281A, 1201–11 (2004); personal communication with Charles Wysocki, September 2006.

In his book The Scented Ape: Stoddart, *The Scented Ape,* 63.

129 *"If armpit odor is a turn-on":* Mel Rosenberg, personal communication, July 29, 2006.

The word "pheromone" . . . was coined: P. Karlson and M. Luscher, "Pheromones: a new term for a class of biologically active substances," *Nature* 183, 55–56 (1959).

sex hormones secreted from the eyes: H. Kimoto, "Sex-specific peptides from exocrine glands stimulate mouse vomeronasal sensory neurons," *Nature* 437, 898–901 (2005).

Among the first clues to the existence: M. K. McClintock, "Menstrual synchrony and suppression," *Nature* 229, 244–45 (1971).

the same effect could be achieved: M. McClintock et al., "Regulation of ovulation by human pheromones," *Nature* 392, 177–79 (1998).

Recently McClintock's team discovered: S. Jacob et al., "Effects of breastfeeding chemosignals on the human menstrual cycle," *Human Reproduction* 19:2, 422–29 (2004); N. A. Spencer, "Social chemosignals from breastfeeding women increase sexual motivation," *Hormones and Behavior* 46, 362–70 (2004).

130 *exposed women to male underarm odors:* G. Preti et al., "Male axillary extracts contain pheromones that affect pulsatile secretion of luteinizing hormone and mood in women recipients," *Biology of Reproduction* 68, 2107–13 (2003).

Scientists asked women to wear a T-shirt: D. Singh and P. M. Bronstad, "Female body odour is a potential cue to ovulation," *Proceedings of the Royal Society of London B* 268, 797–801 (2001).

Lawrence Katz . . . overturned this view: M. Luo et al., "Encoding phero-monal signals in the accessory olfactory bulb of behaving mice," *Science* 299, 1196–1201 (2003).

Since then, several studies have confirmed: S. D. Liberles and L. B. Buck, "A second class of chemosensory receptors in the olfactory epithelium," *Nature* 442, 645–50 (2006); H. Yoon et al., "Olfactory inputs to hypotha-lamic neurons controlling reproduction and fertility," *Cell* 123, 669–82 (2005); Gordon M. Shepherd, "Smells, brains and hormones," *Nature* 439, 149–51 (2006).

131 *Women tend to prefer the odor:* C. Wedekind et al., "MHC-dependent mate preferences in humans," *Proceedings of the Royal Society of London B* 260, 245–49 (1995).

Martha McClintock and colleagues found: S. Jacob et al., "Paternally in-herited MHC alleles are associated with women's choice of male odor," *Nature Genetics* 30, 175–79 (2002).

10. BEWITCHED

133 *"At night, every cat is a leopard":* Giovanni Torriano, *Piazza Universale di Proverbi Italiani; or, A Common Place of Italian Proverbs and Proverbial Phrases* (London, 1666), 171, quoted in A. Roger Ekirch, *At Day's Close: Night in Times Past* (New York: Norton, 2005), 42.

135 *When scientists studied the circadian:* R. Refinetti, "Time for sex: nyc-themeral distribution of human sexual behavior," *Journal of Circadian Rhythms* 3, 4 (2005).

Not surprisingly, research suggests: J. Larson et al., "Morning and night couples: the effect of wake and sleep patterns on marital adjustment," *Journal of Marital and Family Therapy* 17, 53–65 (1991); reported in Mi-chael Smolensky and Lynne Lamberg, *The Body Clock Guide to Better Health* (New York: Holt, 2000), 51.

Levels of testosterone . . . are significantly lower: R. Luboshitzky, "Relation-ship between rapid eye movement sleep and testosterone secretion in normal men," *Journal of Andrology* 20, 731–37 (1999); F. W. Turek, "Bio-logical rhythms in reproductive processes," *Hormone Research* 37 (supp. 3), 93–98 (1992).

Semen quality . . . peaks in the afternoon: A. Cagnacci et al., "Diurnal variation of semen quality in human males," *Human Reproduction* 14:1, 106–9 (1999).

136 *Our understanding of such positive states:* Over the past three decades, some ninety thousand studies have addressed anxiety, anger, and depres-sion, and only five thousand have focused on happiness and joy. Figures from Paul Martin, *Making Happy People* (New York: Harper Perennial, 2006), cited in Maggie McDonald, "Cheer up children," *New Scientist,* February 4, 2006, 56.

Touch is the sense least easily fooled: F. Sachs, "The intimate sense," *The Sciences,* January/February 1988, 28–34.

137 *The purported positive effects of massaging:* Touch Research Institutes, University of Miami School of Medicine, www.miami.edu/touch-research/, retrieved February 23, 2006.

Not long ago, the neurophysiologist Håkan Olausson: H. Olausson, "Unmyelinated tactile afferents signal touch and project to insular cortex," *Nature Neuroscience* 5:9, 900–904 (2002); Olausson, personal communication, September 2006.

138 *Italian researchers probing the hormonal changes:* D. Marazziti and D. Canale, "Hormonal changes when falling in love," *Psychoneuroendocrinology* 29, 931–36 (2004).

Fisher . . . has peered inside the head: H. E. Fisher et al., "Defining the brain systems of lust, romantic attraction, and attachment," *Archives of Sexual Behavior* 31:5, 413–19 (2002); Helen Fisher, personal communication, February 18, 2005.

139 *This result confirmed earlier findings:* A. Bartels and S. Zeki, "The neural basis of romantic love," *Neuroreport* 11:17, 3829–33 (2000).

One widely touted study . . . of Swiss students: M. Kosfeld et al., "Oxytocin increases trust in humans," *Nature* 435, 673–76 (2005).

"A man falls in love through his eyes": Woodrow Wyatt quoted in "Imaging gender differences in sexual arousal," *Nature Neuroscience* 7:4, 325–26 (2004).

studies of sex differences in the processing: S. Hamann et al., "Men and women differ in amygdala response to visual sexual stimuli," *Nature Neuroscience* 7:4, 411–16 (2004).

Women are more sexually aroused: Fisher et al., "Defining the brain systems of lust, romantic attraction, and attachment."

140 *When it comes to mental processing:* D. Kimura, "Sex differences in the brain," www.sciam.com/article.cfm?articleID=00018E9D-879D-1D06-8E49809 EC588EEDF.

Functional MRI studies show: B. A. Shaywitz et al., "Sex differences in the functional organization of the brain for language," *Nature* 373, 607–9 (1995).

In his classic studies on human sexuality: The Kinsey Reports: *Sexual Behavior in the Human Male* (Bloomington: Indiana University Press, 1948, reprint 1998) and *Sexual Behavior in the Human Female* (Philadelphia: Saunders, 1953).

141 *As Leonardo da Vinci wrote:* Leonardo's essay "The Penis" quoted in Serge Bramly, *Leonardo: The Artist and the Man,* trans. Sian Reynolds (London: Edward Burlingame Books, 1991).

Contributing to the blood flow: K. J. Hurt et al., "Akt-dependent phosphorylation of endothelial nitric-oxide synthase mediates penile erection," *Proceedings of the National Academy of Sciences* 99:6, 4061–66 (2002).

A study by Lique Coolen: W. A. Truitt and L. M. Coolen, "Identification of a potential ejaculation generator in the spinal cord," *Science* 297, 1566–69 (2002).

142 *Work by Coolen:* L. M. Coolen et al., "Activation of mu opioid receptors in the medial preoptic area following copulation in male rats," *Neuroscience* 124:1, 11–21 (2003).

As for women: A. M. Traish et al., "Biochemical and physiological mechanisms of female genital sexual arousal," *Archives of Sexual Behavior* 31:5, 393–400 (2002).

Yes, the G spot is real: M. Giorgi et al., "Type 5 phosphodiesterase expression in the human vagina," *Urology* 60, 191–95 (2002).

Gentle pressure on the spot: B. Whipple and B. R. Komisaruk, "Elevation of pain threshold by vaginal stimulation in women," *Pain* 21, 357–67 (1985); B. Whipple and B. R. Komisaruk, "Analgesia produced in women by genital self-stimulation," *Journal of Sex Research* 24:1, 130–40 (1988).

British researchers found that having sex: S. Brody, "Blood pressure reactivity to stress is better for people who recently had penile-vaginal intercourse than for people who had other or no sexual activity," *Biological Psychology* 71, 214–22 (2006).

scientists at Rutgers University studying women: B. R. Komisaruk et al., "Brain activation during vaginocervical self-stimulation and orgasm in women with complete spinal cord injury: fMRI evidence of mediation by the vagus nerves," *Brain Research* 1024, 77–88 (2004).

One new study points an intriguing finger: K. Dunn et al., "Genetic influences on variation in female orgasmic function: a twin study," *Biology Letters,* June 2005; online edition, DOI: 10.1098/rsbl.2005.0308.

143 *A case report in the* Lancet: P. J. Reading and R. G. Will, "Unwelcome orgasms," *Lancet* 350, 1746 (1997).

Orgasm is . . . a cerebral experience: J. P. Changeux, *Neuronal Man: The Biology of the Mind* (Princeton, N.J.: Princeton University Press, 1997), 112–14.

Dutch scientists shocked the neuroscience community: G. Holstege et al., "Brain activation during human male ejaculation," *Journal of Neuroscience* 23, 9185–93 (2003); J. R. Georgiadis et al., "Brain activation during female sexual orgasm," *Society of Neuroscience Abstracts* 727:7, 31 (2003); J. R. Georgiadis et al., "Deactivation of the amygdala during human male sexual behavior," program no. 727.6, Society for Neuroscience meeting, November 8–12, 2003, New Orleans; B. R. Komisaruk and B. Whipple, "Functional MRI of the brain during orgasm in women," *Annual Review of Sex Research* 16, 62–86 (2005).

144 *men who reported the highest frequency:* G. D. Smith et al., "Sex and death: are they related? Findings from the Caerphilly cohort study," *British Medical Journal* 315, 1641–44 (1997); S. Ebrahim et al., "Sexual intercourse and risk of ischaemic stroke and coronary heart disease: the Caerphilly study," *Journal of Epidemiology and Community Health* 56, 99–102 (2002).

college students who have sexual intercourse: C. J. Charnetski and F. X.

Brennan, "Sexual frequency and immunoglobulin A (IgA)," paper presented at the annual meeting of the Eastern Psychological Association, Providence, R.I., 1999.

144 *In a sample of sexually active college females:* G. Gallup et al., "Does semen have antidepressant properties?," *Archives of Sexual Behavior* 31:3, 289–93 (2002).

The scientists are quick to say: R. Persaud, "Semen acts as an anti-depressant," *New Scientist,* June 26, 2002, www.newscientist.com/article. ns?id=dn2457.

11. NIGHT AIRS

145 *A sixteenth-century Italian priest:* Quoted in A. Roger Ekirch, *At Day's Close* (New York: Norton, 2005), 13.

many ills worsen at night: The following discussion of circadian aspects of disease comes from M. H. Smolensky and M. L. Bing, "Chronobiology and chronotherapeutics in primary care," *Patient Care* (Clinical Focus supp.), Summer 1997, 1–21; M. H. Smolensky et al., "Medical chronobiology: concepts and applications," *American Review of Respiratory Disease* 147:6 (part 2), S2–19; Michael Smolensky and Lynne Lamberg, *The Body Clock Guide to Better Health* (New York: Holt, 2000); Russell Foster and Leon Kreitzman, *Rhythms of Life* (London: Profile Books, 2004), 212 f; G. A. Bjarnason and R. Jordan, "Rhythms in human gastrointestinal mucosa and skin," *Chronobiology International* 19:1, 129–40 (2002).

asthma attacks hundreds of times: Foster and Kreitzman, *Rhythms of Life,* 224; R. J. Martin, "Small airway and alveolar tissue changes in nocturnal asthma," *American Journal of Respiratory and Critical Care Medicine* 157:5, S188–90 (1998).

bronchial passageways that move air: Martin, "Small airway and alveolar tissue changes in nocturnal asthma."

146 *adults get two to four colds a year:* F. Hayden, "Introduction: emerging importance of the rhinovirus," *American Journal of Medicine* 112:6A, 1s–3s (2002); J. M. Gwaltney, "Rhinoviruses," in A. S. Evans and R. A. Kaslow, eds., *Viral Infection of Humans: Epidemiology and Control,* 4th ed. (New York: Plenum Press, 1997), 815–38.

Researchers have carefully computed: A. M. Fendrick, "The economic burden of non-influenza-related viral respiratory tract infection in the United States," *Archives of Internal Medicine* 163:4, 487–94 (2003).

The Greek philosopher Celsus wrote: Celsus, *De Medicina,* vol. 2, ed. W. G. Spencer (London: W. Heinemann, 1938), 91.

Scientists persuaded one group: H. F. Dowling et al., "Transmission of the common cold to volunteers under controlled conditions," *American Journal of Hygiene* 68, 659–65 (1958).

A decade later, a similar experiment: R. G. Douglas et al., "Exposure to cold environment and rhinovirus common cold: failure to demonstrate effect," *New England Journal of Medicine* 279, 742–47 (1968).

147   *a new study offers some evidence:* C. Johnson and R. Eccles, "Acute cooling of the feet and the onset of common cold symptoms," *Family Practice* 22:6, 608–13 (2005).
However, say skeptics: personal communication with J. Owen Hendley, February 2007.
That colds tend to flourish: The following discussion of colds and cold viruses comes from an interview with Jack Gwaltney, March 8, 2004; J. M. Gwaltney, "Viral respiratory infection therapy: historical perspectives and current trials," *American Journal of Medicine* 112:6A, 33s–41s (2002).
Viruses are highly contagious bugs: J. M. Gwaltney, "Clinical significance and pathogenesis of viral respiratory infections," *American Journal of Medicine* 112:6A, 13s–18s (2002). J. M. Harris and J. M. Gwaltney, "Incubation periods of experimental rhinovirus infection and illness," *Clinical Infectious Diseases* 23, 1287–90 (1996).
rhinoviruses survive and remain infectious: J. M. Gwaltney and J. O. Hendley, "Rhinovirus transmission: one if by air, two if by hand," *American Journal of Epidemiology* 107, 357–61 (1978).
During only ten seconds of hand exposure: J. M. Gwaltney et al., "Hand-to-hand transmission of rhinovirus colds," *Annals of Internal Medicine* 88:4, 463–67 (1978).

148   *transmission could be interrupted:* J. M. Gwaltney, "Transmission of experimental rhinovirus infection by contaminated surfaces," *American Journal of Epidemiology* 116:5, 828–33 (1982); Arnold Monto, "Epidemiology of viral respiratory infections," *American Journal of Medicine* 112:6A, 4s–12s (2002).
These structures normally swell: Donald Proctor and Ib Andersen, eds., *The Nose: Upper Airway Physiology and the Atmospheric Environment* (New York: Elsevier Biomedical Press, 1982), 203.
forceful nose-blowing may do: J. M. Gwaltney et al., "Nose blowing propels nasal fluid into the paranasal sinuses," *Clinical Infectious Diseases* 30, 387–91 (2000).
If the tickle of mucus sufficiently: L. Suranyi, "Localization of the 'sneeze center,'" *Neurology* 57:1, 161 (2001).

149   *Coughing can achieve even greater powers:* R. S. Irwin et al., "Managing cough as a defense mechanism and a symptom: a consensus panel report of the American College of Chest Physicians," *Chest* 114 (supp. 2), 133s–81s (1998), www.chestjournal.org/cgi/reprint/114/2/133S.pdf.
Once considered a simple reflex: S. B. Mazzone, "An overview of the sensory receptors regulating cough," *Cough* 1:2, DOI: 10.1186/1745-9974-1-2 (2005); J. G. Widdicombe, "Afferent receptors in the airways and cough," *Respiratory Physiology* 114, 5–15 (1998); S. B. Mazzone, "Sensory regulation of the cough reflex," *Pulmonary Pharmacology and Therapy* 17, 361–68 (2004).
Runny nose, sneezing, cough: The following discussion is from an interview with Gwaltney, March 8, 2004; B. Winther et al., "Viral-induced rhinitis," *American Journal of Rhinology* 12:1, 17–20 (1998).

150 *chronic stress, which has been linked:* S. Cohen et al., "Types of stressors that increase susceptibility to the common cold in healthy adults," *Health Psychology* 17:3, 214–23 (1998); J. M. Gwaltney and F. G. Hayden, "Psychological stress and the common cold," *New England Journal of Medicine* 325, 644 (1992).

not everyone exposed to a virus: J. M. Gwaltney, "Clinical significance and pathogenesis of viral respiratory infections," *American Journal of Medicine* 112:6A, 13s–18s (2002).

151 *Antihistamines suppress sneezing:* P. S. Muether and J. M. Gwaltney, "Variant effect of first- and second-generation antihistamines as clues to their mechanism of action on the sneeze reflex in the common cold," *Clinical Infectious Diseases* 33, 1483–88 (2001).

A major review of nonprescription cough medicines: Knut Schroeder and Tom Fahey, "Systematic review of randomised controlled trials of over the counter cough medicines for acute cough in adults," *British Medical Journal* 324, 329 (2002).

A century later, a popular remedy: Scientific American, May 1895, quoted in *Scientific American,* May 1995, 10.

Gwaltney has been working: J. M. Gwaltney et al., "Combined antiviral-antimediator treatment for the common cold," *Journal of Infectious Diseases* 186, 147–54 (2002); J. M. Gwaltney, "Viral respiratory infection therapy: historical perspectives and current trials," *American Journal of Medicine* 112:6A, 33s–41s (2002).

152 *Graphs of sneezing, stuffy nose:* A. C. Grant and E. P. Roter, "Circadian sneezing," *Neurology* 44:3, 369–75 (1994).

Cough frequency, too, shows a marked: J. Kuhn et al., "Antitussive effect of guaifenesin in young adults with natural colds," *Chest* 82:6, 713–18 (1982).

Illnesses of many types are affected: M. H. Smolensky et al., "Medical chronobiology: concepts and applications," *American Review of Respiratory Disease* 147:6 (part 2), S2–S19 (1993); Smolensky and Lamberg, *The Body Clock Guide to Better Health;* Foster and Kreitzman, *Rhythms of Life,* 212 f.

But as Smolensky points out: Smolensky et al., "Medical chronobiology: concepts and applications."

A patient may be diagnosed as normal: Y. Watanabe et al., "Thousands of blood pressure and heart rate measurements at fixed clock hours may mislead," *Neuroendocrinology Letters* 24:5,339–40 (2003).

surveys suggest that physicians: M. H. Smolensky, "Knowledge and attitudes of American physicians and public about medical chronobiology and chronotherapeutics. Findings of two 1996 Gallup surveys." *Chronobiology International* 15, 377–94 (1998); Foster and Kreitzman, *Rhythms of Life,* 226.

153 *Though direct evidence of circadian modulation:* C. B. Green and J. S. Takahashi, "Xenobiotic metabolism in the fourth dimension: PARtners

in time," *Cell Metabolism* 4:1, 3–4 (2006). Personal communication with Carla B. Green, October 2006.

Lidocaine . . . relieved dental pain: A. Reinberg and M. Reinberg, "Circadian changes of the duration of action of local anaesthetic agents," *Naunyn-Schmiedeberg's Archives of Pharmacology* 297, 149–59 (1977).

On the other hand, a 2006 report showed: M. C. Wright et al., "Time of day effects on the incidence of anesthetic adverse events," *Quality and Safety in Health Care* 15:4, 258–63 (2006).

Such time-of-day effects have been documented: G. A. Bjarnason et al., "Circadian variation in the expression of cell-cycle proteins in human oral epithelium," *American Journal of Pathology* 154, 613–22 (1999).

The goal . . . should be to balance: Foster and Kreitzman, *Rhythms of Life*, 215.

154 *Many anticancer drugs:* Smolensky and Lamberg, *The Body Clock Guide to Better Health*, 227–29. G. A. Bjarnason and R. Jordan, "Rhythms in human gastrointestinal mucosa and skin," *Chronobiology International* 19:1, 129–40 (2002); Foster and Kreitzman, *Rhythms of Life*, 216–19.

Francis Lévi believes: The following discussion of Lévi's work comes from his "Circadian interactions with cancer," presented at the Society for Research on Biological Rhythms annual meeting, Amelia Island, Florida, 2002; M. C. Mormont and F. Lévi, "Cancer chronotherapy: principles, applications, and perspectives," *Cancer* 98:4, 881–82 (2003).

William Hrushesky . . . has found that the cells: K. Buchi et al., "Circadian rhythm of cellular proliferation in the human rectal mucosa," *Gastroenterology* 101, 410–15 (1991).

155 *Hrushesky published a study on the timing:* W. Hrushesky, "Circadian timing of cancer chemotherapy," *Science* 228, 73–75 (1985). Similar results have been found by researchers studying childhood leukemia. In a trial involving 118 children with acute leukemia, those who received medications in the late afternoon or evening were three times more likely to have had their cancer go into remission than those treated in the morning. G. E. Rivard et al., "Circadian time-dependent response of childhood lymphoblastic leukemia to chemotherapy: a long-term follow-up study of survival," *Chronobiology International* 10, 201–4 (1993).

Francis Lévi has had similar success: F. Lévi et al., "Chronotherapy of colorectal cancer metastases," *Hepatogastroenterology* 48, 320–22 (2001).

12. SLEEP

157 *"Sleep is the most moronic":* Vladimir Nabokov, *Speak, Memory* (New York: Vintage, 1989), 108.

The brain is hardly "absent": The following material on sleep is from Jerome M. Siegel, "The phylogeny of sleep," presented at the Society for Research on Biological Rhythms annual meeting, Amelia Island, Florida,

2002 (hereafter, SRBR meeting, 2002); personal communication with Jerome Siegel, February 15, 2005; J. M. Siegel, "Clues to the functions of mammalian sleep," *Nature* 437, 1264–71 (2005).

158   *Even during deep sleep:* J. A. Hobson, "Sleep is of the brain, by the brain, and for the brain," *Nature* 437, 1254 (2005).

In maintaining good health: William C. Dement and Christopher Vaughan, *The Promise of Sleep* (New York: Dell, 2000).

The shift, it turns out, is executed: Quotes from Saper referring to the sleep switch are from C. B. Saper, "Hypothalamic regulation of sleep and circadian rhythms," *Nature* 437, 1257–63 (2005).

Baron Constantin von Economo . . . first identified: C. von Economo, "Sleep as a problem of localization," *Journal of Nervous and Mental Disorders* 71, 249–59 (1930).

Scientists have recently pinpointed the switch: Saper, "Hypothalamic regulation of sleep and circadian rhythms."

159   *But scientists suspect that the switch:* C. S. Colwell and S. Michel, "Sleep and circadian rhythms: do sleep centers talk back to the clock?," *Nature Neuroscience* 6:10, 1005–6 (2003); T. Deboer et al., "Sleep states alter activity of suprachiasmatic nucleus neurons," *Nature Neuroscience* 6:10, 1086–90 (2003); D. J. Dijk and C. A. Czeisler, "Contribution of the circadian pacemaker and the sleep homeostat to sleep propensity, sleep structure, electroencephalographic slow waves, and sleep spindle activity in humans," *Journal of Neuroscience* 15, 3526–38 (1995).

When sleep seized my daughter: The following description of sleep stages draws from Dement and Vaughan, *The Promise of Sleep*, 18–22; Peretz Lavie, *The Enchanted World of Sleep* (New Haven: Yale University Press, 1996), 26 f; J. M. Siegel, "Why we sleep," *Scientific American*, November 2003, 92–97.

160   *When scientists implanted tiny sensors:* K. J. Noonan, "Growing pains: are they due to increased growth during recumbency as documented in a lamb model?," *Journal of Pediatric Orthopaedics* 24, 6 (2004).

When Swiss researchers studied a gene: J. V. Rétey et al., "A functional genetic variation of adenosine deaminase affects the duration and intensity of deep sleep in humans," *Proceedings of the National Academy of Sciences* 102, 15676–81 (2005).

but a complex system of neurotransmitters: J. M. Siegel, "Clues to the functions of mammalian sleep," *Nature* 437, 1264–71 (2005).

161   *Everyone dreams, says J. Allan Hobson:* Hobson, "Sleep is of the brain, by the brain, and for the brain."

Recent advances in imaging: Siegel, "Why we sleep."

Lively are the cortical regions: T. A. Nielsen and P. Stenstrom, "What are the memory sources of dreaming?," *Nature* 437, 1286–89 (2005).

nightmares may in some cases arise: T. Nielsen, "Chronobiological features of dream production," *Sleep Medicine Reviews* 8, 403–24 (2004).

162   *women typically have more nightmares:* T. A. Nielsen, "The typical dreams of Canadian university students," *Dreaming* 13:4, 211 (2003).

"This gender difference is quite robust": Tore Nielsen, personal communication, September 2006.

"REM dreams tend to be forgotten": A. Rechtschaffen, personal communication, February 11, 2005.

163 *but it does tend to improve sleep patterns:* S. D. Youngstedt and C. E. Kline, "Epidemiology of exercise and sleep," *Sleep and Biological Rhythms* 4:3, 215 (2006); see also S. D. Youngstedt et al., "No association of sleep with total daily physical activity in normal sleepers," *Physiology and Behavior* 78, 395–401 (2003); S. D. Youngstedt, "Does exercise truly enhance sleep?," *Physician and Sports Medicine* 25:10, 72–82 (1997).

Researchers recently glimpsed why: J. Mu et al., "Ethanol influences on native T-type calcium current in thalamic sleep circuitry," *Journal of Pharmacology and Experimental Therapy* 307:1, 197–204 (2003).

164 *"If sleep does not serve":* A. Rechtschaffen and B. M. Bergmann, "Sleep deprivation in the rat: an update of the 1989 paper," *Sleep* 25, 18–24 (2002).

William Dement, who observed him: Dement and Vaughan, *The Promise of Sleep,* 245.

"There's another long-term, massive experiment": This and the following quotes from Charles Czeisler are from C. Czeisler, "Sleep: what happens when doctors do without it?," Medical Center Hour, University of Virginia, March 1, 2006.

A 2005 poll taken by the National Sleep Foundation: www.sleepfoundation.org/press/index.php?secid=&id=120.

165 *As recently as a decade ago, eminent researchers:* Lavie, *The Enchanted World of Sleep,* 114; C. A. Czeisler, "Quantifying consequences of chronic sleep restriction," *Sleep* 26:3, 247–48 (2003); H.P.A. Van Dongen et al., "The cumulative cost of additional wakefulness," *Sleep* 26:2, 117 (2003).

166 *just how much sleep is best:* See D. F. Kripke et al., "Mortality associated with sleep duration and insomnia," *Archives of General Psychiatry* 59, 131–36 (2002).

people who slept less than six hours a night: Van Dongen et al., "The cumulative cost of additional wakefulness."

in terms of sedative effects: T. Roehrs et al., "Ethanol and sleep loss: a 'dose' comparison of impairing effects," *Sleep* 26:8, 981–85 (2003); D. Dawson and K. Reid, "Fatigue, alcohol and performance impairment," *Nature* 388, 235 (1997).

"With one night of short sleep": Czeisler, "Sleep: what happens when doctors do without it?"; L. K. Barger et al., "Extended work shifts and the risk of motor vehicle crashes among interns," *New England Journal of Medicine* 352, 125–34 (2005).

When you're driving . . . at sixty: Dement and Vaughan, *The Promise of Sleep,* 218.

drowsiness increases a driver's risk: "Breakthrough research on real-world driver behavior released," press release, April 20, 2006, www.nhtsa.gov; editorial page, *Nature Insight: Sleep* 437, 1206 (2005).

167 *As William Dement points out:* Dement and Vaughan, *The Promise of Sleep,* 51–53; Russell Foster and Leon Kreitzman, *Rhythms of Life* (London: Profile Books, 2004), 208–9.
The report on the disaster: http://history.nasa.gov/rogersrep/v2appg.htm #g25.
scientists relied on subjects' self-reported: Czeisler, "Quantifying consequences of chronic sleep restriction."
To parse the popular belief: K. Spiegel et al., "Effect of sleep deprivation on response to immunization," *Journal of the American Medical Association* 288:12, 1471–72 (2002).
Van Cauter also found that sleep loss: K. Spiegel et al., "Impact of sleep debt on metabolic and endocrine function," *Lancet* 354, 1435–39 (1999); E. Tasali, "Slow wave activity levels are correlated with insulin secretion in healthy young adults," *Sleep* 26 (abstract supp.), A62 (2003).

168 *She and her team have reported:* K. Spiegel et al., "Brief communication: sleep curtailment in healthy young men is associated with decreased leptin levels, elevated ghrelin levels, and increased hunger and appetite," *Annals of Internal Medicine* 141, 846–50 (2004); see also K. Spiegel, "Sleep curtailment results in decreased leptin levels and increased hunger and appetite," *Sleep* 26 (abstract supp.), A174 (2003).
obesity is tightly correlated: J. E. Gangwisch et al., "Inadequate sleep as a risk factor for obesity: analyses of the NHANES I," *Sleep* 28:10, 1217–20 (2005). For a similar study of children, see J.-P. Chaput et al., "Relationship between short sleeping hours and childhood overweight/obesity: results from the 'Quebec en Forme' Project," *International Journal of Obesity,* 30:7, 1080–85 (2006).
A twist to this tale comes: report presented by Sanjay Patel at the American Thoracic Society international conference, May 23, 2006, San Diego.

169 *"The effects of sleep deprivation alone":* A. Rechtschaffen and B. M. Bergmann, "Sleep deprivation in the rat: an update of the 1989 paper," *Sleep* 25, 18–24 (2002); personal communication with Rechtschaffen, February 16, 2005.
Jerry Siegel and others have delved: The following discussion comes from J. M. Siegel, "Clues to the functions of mammalian sleep," *Nature* 437, 1264–71 (2005); Siegel, "The phylogeny of sleep"; Siegel, "Why we sleep."

170 *Siegel suspects that one key:* Siegel, "Clues to the functions of mammalian sleep"; J. M. Siegel and M. A. Rogawski, "A function for REM sleep: regulation of noradrenergic receptor sensitivity," *Brain Research Review* 13, 213–33 (1988).

171 REM *steps in to keep neurons:* M. J. Morrissey, "Paradoxical sleep and its role in the prevention of apoptosis in the developing brain," *Sleep* 26 (abstract supp.), A46 (2003).
a group of Finnish researchers: M. Cheour et al., "Speech sounds learned by sleeping newborns," *Nature* 415, 599–600 (2002).

172 *Robert Stickgold and his colleagues:* R. Stickgold et al., "Visual discrimina-

tion learning requires sleep after training," *Nature Neuroscience* 3, 1237–38 (2000); M. P. Walker et al., "Practice with sleep makes perfect: sleep-dependent motor skill learning," *Neuron* 35, 205–11 (2002).

Some years ago, scientists found: P. Maquet et al., "Experience-dependent changes in cerebral activation during REM sleep," *Nature Neuroscience* 3:8, 831–36 (2000).

One recent study showed that the precise: P. Peigneux et al., "Are spatial memories strengthened in the human hippocampus during slow-wave sleep?," *Neuron* 44, 535–45 (2004).

Perhaps sleep allows the brain: I. S. Hairston and R. R. Knight, "Sleep on it," *Nature* 430, 27–28 (2004).

Tononi and his colleagues reported: R. Huber et al., "Local sleep and learning," *Nature* 430, 78–81 (2004).

A team led by Ullrich Wagner: U. Wagner et al., "Sleep inspires insight," *Nature* 427, 352–55 (2004).

173 *The annals of literature:* Lavie, *The Enchanted World of Sleep,* 90.

Robert Louis Stevenson claimed: Lavie, *The Enchanted World of Sleep,* 90; James Pope Hennessy, *Robert Louis Stevenson* (New York: Simon and Schuster, 1975), 207; quote from Stevenson, *Across the Plains,* http://sunsite.berkeley.edu/literature/stevenson/plains/plains8.html.

Science also holds stories: Lavie, *The Enchanted World of Sleep,* 90; Paolo Mazzarello, "What dreams may come," *Nature* 408, 523 (2000); Dement and Vaughan, *The Promise of Sleep,* 321.

13. HOUR OF THE WOLF

175 *When William Dement and Mary Carskadon studied:* M. A. Carskadon et al., "Sleep and daytime sleepiness in the elderly," *Journal of Geriatric Psychiatry* 13:2, 135–51 (1980); R. M. Coleman et al., "Sleep-wake disorders in the elderly: polysomnographic analysis," *Journal of the American Geriatric Society* 27:9, 289–96 (1981); William C. Dement and Christopher Vaughan, *The Promise of Sleep* (New York: Dell, 2000), 121; see also D. J. Dijk et al., "Age-related increase in awakenings: impaired consolidation of non-REM sleep at all circadian phases," *Sleep* 24:5, 565–77 (2001).

This loss of deep sleep actually begins in midlife: E. Van Cauter, "Age-related changes in slow wave sleep and REM sleep and relationship with growth hormone and cortisol levels in healthy men," *Journal of the American Medical Association* 284, 861–68 (2000).

Some evidence suggests that in the elderly: J. F. Duffy et al., "Later endogenous circadian temperature nadir relative to an earlier wake time in older people," *American Journal of Physiology* 275:5 (part 2), R1478–87 (1998); E. Van Cauter et al., "Effects of gender and age on the levels of circadian rhythmicity of plasma cortisol," *Journal of Clinical Endocrinology and Metabolism* 81, 2468–73 (1996); T. Reilly et al., "Aging, rhythms of physical performance, and adjustment to changes in the sleep-activity cycle," *Occupational and Environmental Medicine* 54, 812–16 (1997); J. F. Duffy and C. A. Czeisler, "Age-related change in the relationship between circadian period, circadian phase, and diurnal preference in humans," *Neuroscience Letters* 318:3, 117–20 (2002); C. A. Czeisler et al., "Association of sleep-wake habits in older people with changes in output of circadian pacemaker," *Lancet* 340, 933–36 (1992); F. Aujard et al., "Circadian rhythms in firing rate of individual suprachiasmatic nucleus neurons from adult and middle-aged mice," *Neuroscience* 106:2, 255–61 (2001); E. Satinoff, "Patterns of circadian body temperature rhythms in aged rats," *Clinical and Experimental Pharmacology and Physiology* 25:2, 135–40 (1998).

176   *They may in part be rooted:* W. N. Charman, "Age, lens transmittance, and the possible effects of light on melatonin suppression," *Ophthalmic and Physiological Optics* 23, 181–87 (2003).

   Scientists know that normal aging: M. D. Madeira et al., "Age and sex do not affect the volume, cell numbers, or cell size of the suprachiasmatic nucleus of the rat: an unbiased stereological study," *Journal of Comparative Neurology* 361:4, 585–601 (1995).

   But at least one new study . . . suggests that aging: S. Yamazaki, "Effects of aging on central and peripheral mammalian clocks," *Proceedings of the National Academy of Sciences* 99:16, 10801–6 (2002); F. Aujard et al., "Circadian rhythms in firing rate"; D. E. Kolker, "Aging alters circadian and light-induced expression of clock genes in golden hamsters," *Journal of Biological Rhythms* 18:2, 159–69 (2003).

   In traditional, non-Western societies: C. M. Worthman and M. Melby, "Toward a comparative developmental ecology of human sleep," in M. A. Carskadon, ed., *Adolescent Sleep Patterns: Biological, Social, and Psychological Influences* (New York: Cambridge University Press), 69–117; personal communication with Carol Worthman, August 8, 2006.

177   *Not much was known about past patterns:* A. Roger Ekirch, *At Day's Close* (New York: Norton, 2005).

   The habit of sleeping in bouts: I. Tobler, "Napping and polyphasic sleep in mammals," in D. F. Dinges and R. J. Broughton, eds., *Sleep and Alertness: Chronobiological, Behavioral, and Medical Aspects of Napping* (New York: Raven Press, 1989), 9–30.

   Thomas Wehr . . . once devised an experiment: T. A. Wehr, "In short photoperiods, human sleep is biphasic," *Journal of Sleep Research* 1:2, 103–7 (1992).

178   *scientists at Vanderbilt University showed:* H. Ohta et al., "Constant light desynchronizes mammalian clock neurons," *Nature Neuroscience* 8:3, 267–69 (2005).

179   *In turning on lamps and lights:* J. M. Zeitzer et al., "Temporal dynamics of late-night photic stimulation of the human circadian timing system," *American Journal of Physiology: Regulatory, Integrative, and Comparative Physiology* 289:3, R839–44 (2005).

Exposure to even low light levels: D. B. Boivin et al., "Dose-response relationships for resetting of human circadian clock by light," *Nature* 379, 540–42 (1996).

Charles Czeisler's team has found: S. B. S. Khalsa et al., "A phase response curve to single bright light pulses in human subjects," *Journal of Physiology* 549 (Pt. 3): 945–52 (2003); J. M. Zeitzer et al., "Sensitivity of the human circadian pacemaker to nocturnal light: melatonin phase resetting and suppression," *Journal of Physiology* 526:3, 695–702 (2000).

Even brief exposure to light: J. A. Gastel, "Melatonin production: proteasomal proteolysis in serotonin N-acetyltransferase regulation," *Science* 279, 1358–60 (1998).

When you fly halfway around the world: C. Dunlop and J. Cortazar in *Los Autonautas de la Cosmopista o un Viage Atemporal* (1983), quoted in Russell Foster and Leon Kreitzman, *Rhythms of Life* (London: Profile Books, 2004), 201.

180 *In one study, Menaker and his colleagues:* S. Yamazaki et al., "Resetting central and peripheral circadian oscillators in transgenic rats," *Science* 288, 682 (2000); personal communication with Michael Menaker, March 2005.

inspired by his own jet lag symptoms: Kwangwook Cho, "Chronic 'jet lag' produces temporal lobe atrophy and spatial cognitive deficits," *Nature Neuroscience* 4:6, 567–68 (2001); K. Cho et al., "Chronic jet lag produces cognitive deficits," *Journal of Neuroscience* 20, RC66 (2000).

181 *some 15 percent of the U.S. workforce:* Bureau of Labor Statistics, "Workers on flexible and shift schedules in 2004 summary," www.bls.gov/news.release/flex.nro.htm, retrieved October 16, 2006.

To flesh out the effects: J. Arendt, "Shift-work: adapting to life in a new millennium," presentation at the 2002 meeting of the Society for Sleep Research and Biological Rhythms, Amelia Island, Florida; personal communication with Josephine Arendt, March 21, 2005.

182 *the 78,500 women in the Nurses' Health Study:* E. S. Schernhammer et al., "Rotating night shifts and risk of breast cancer in women participating in the Nurses' Health Study," *Journal of the National Cancer Institute* 93:20, 1563–68 (2001); E. S. Schernhammer et al., "Night-shift work and risk of colorectal cancer in the Nurses' Health Study," *Journal of the National Cancer Institute* 95:11, 825–28 (2003).

Japanese study of more than 14,000 men: T. Kubo et al., "Prospective cohort study of the risk of prostate cancer among rotating-shift workers: findings from the Japan collaborative cohort study," *American Journal of Epidemiology* 164:6, 549–55 (2006).

researchers have tampered with: L. Fu et al., "The circadian gene *Period2* plays an important role in tumor suppression and DNA-damage response in vivo," *Cell* 111, 41–50 (2002); M. Rosbash and J. S. Takahashi, "The cancer connection," *Nature* 420, 373–74 (2002).

In a 2006 study, William Hrushesky: P. A. Wood et al., "Circadian clock

BMAL-1 nuclear translocation gates WEE1 coordinating cell cycle progression, thymidylate synthase, and 5-fluorouracil therapeutic index," *Molecular Cancer Therapeutics* 5:8, 2023–33 (2006).

182 *But the first strong experimental evidence:* D. E. Blask et al., "Melatonin-depleted blood from premenopausal women exposed to light at night stimulates growth of human breast cancer xenografts in nude rats," *Cancer Research* 65, 11174–84 (2005).

183 *crisis at the Three Mile Island nuclear plant: Report of the President's Commission on the Accident at Three Mile Island* (Washington, D.C.: U.S. Government Printing Office, 1979), www.pddoc.com/tmi2/kemeny/accident/htm.

And the world's worst nuclear accident: M. A. Anderson, "Living in the shadow of Chernobyl," *Science* 292, 420–21 (2001).

The tradition of long work hours: C. Czeisler, "Sleep: what happens when doctors do without it?," Medical Center Hour, University of Virginia, March 1, 2006; Howard Markel, "The accidental addict," *New England Journal of Medicine* 352, 966–68 (2005).

184 *a team of scientists at the Harvard Work Hours:* C. Landrigan et al., "Effect of reducing interns' work hours on serious medical errors in intensive care units," *New England Journal of Medicine* 351, 1838–48 (2004).

"When people have been awake": This and the following quotes are from Czeisler, "Sleep: what happens when doctors do without it?"

A 2006 study by the Harvard group: N. Ayas et al., "Extended work duration and the risk of self-reported percutaneous injuries in interns," *Journal of the American Medical Association* 296, 1055–62 (2006).

Studies of people who sleep: J. K. Wyatt et al., "Circadian temperature and melatonin rhythms, sleep, and neurobehavioral function in humans living on a 20-h day," *American Journal of Physiology* 277:4 (part 2), R1152–63 (1999).

In 2005, Czeisler's group reported: L. K. Barger et al., "Extended work shifts and the risk of motor vehicle crashes among interns," *New England Journal of Medicine* 352, 125–34 (2005).

185 *"In the face of this evidence":* personal communication, Christopher Landrigan, October 2006.

a study conducted by Landrigan: C. P. Landrigan et al., "Interns' compliance with accreditation council for graduate medical education workhour limits," *Journal of the American Medical Association* 296:9, 1063–70 (2006).

There are, in fact, such "lifestyle" drugs: J. K. Walsh, "Modafinil improves alertness, vigilance, and executive function during simulated night shifts," *Sleep* 27:3, 434–39 (2004).

they're not always entirely effective: C. A. Czeisler et al., "Modafinil for excessive sleepiness associated with shift-work sleep disorder," *New England Journal of Medicine* 353:5, 476–86 (2005).

186 *In a few hours will come the peak hour:* Dement and Vaughan, *The Prom-*

ise of Sleep, 107; Foster and Kreitzman, *Rhythms of Life,* 12; Michael Smolensky and Lynne Lamberg, *The Body Clock Guide to Better Health* (New York: Holt, 2000), 133.

"The gods confound the man": Quoted in Foster and Kreitzman, *Rhythms of Life,* 12.

scientists devised an "optical lattice" clock: M. Takamoto et al., "An optical lattice clock," *Nature* 435, 321–24 (2005).

187   *when the problem in life:* Letter to Gerald Brennan, December 25, 1922, in Nigel Nicolson and Joanne Trautmann, eds., *The Letters of Virginia Woolf,* vol. 2 (New York: Harcourt, 1976), 598.

What may lie at the heart: L. Shearman et al., "Interacting molecular loops in the mammalian circadian clock," *Science* 288, 1013–19 (2000).

Index